REVISE EDEXCEL GCSE (9–1)
Combined Science
REVISION WORKBOOK
Higher

Series Consultant: Harry Smith
Authors: Stephen Hoare, Nigel Saunders, Catherine Wilson

Notes from the publisher

1. In order to ensure that this resource offers high-quality support for the associated Pearson qualification, it has been through a review process by the awarding body. This process confirms that this resource fully covers the teaching and learning content of the specification or part of a specification at which it is aimed. It also confirms that it demonstrates an appropriate balance between the development of subject skills, knowledge and understanding, in addition to preparation for assessment.

Endorsement does not cover any guidance on assessment activities or processes (e.g. practice questions or advice on how to answer assessment questions) included in the resource, nor does it prescribe any particular approach to the teaching or delivery of a related course.

While the publishers have made every attempt to ensure that advice on the qualification and its assessment is accurate, the official specification and associated assessment guidance materials are the only authoritative source of information and should always be referred to for definitive guidance.

Pearson examiners have not contributed to any sections in this resource relevant to examination papers for which they have responsibility.

Examiners will not use endorsed resources as a source of material for any assessment set by Pearson.

Endorsement of a resource does not mean that the resource is required to achieve this Pearson qualification, nor does it mean that it is the only suitable material available to support the qualification, and any resource lists produced by the awarding body shall include this and other appropriate resources.

2. Pearson has robust editorial processes, including answer and fact checks, to ensure the accuracy of the content in this publication, and every effort is made to ensure this publication is free of errors. We are, however, only human, and occasionally errors do occur. Pearson is not liable for any misunderstandings that arise as a result of errors in this publication, but it is our priority to ensure that the content is accurate. If you spot an error, please do contact us at resourcescorrections@pearson.com so we can make sure it is corrected.

For the full range of Pearson revision titles across KS2, KS3, GCSE, Functional Skills, AS/A Level and BTEC visit:
www.pearsonschools.co.uk/revise

Contents

Edexcel publishes Sample Assessment
Material and the Specification on its
website. This is the official content and
this book should be used in conjunction
with it. The questions have been written to
help you practise every topic in the book.
Remember: the real exam questions may
not look like this.

Plant and animal cells

1 (a) Which of the following are found in both animal and plant cells?

☐ **A** cell membrane, nucleus, chloroplast

☑ **B** cell membrane, nucleus, ribosomes

☐ **C** cell wall, nucleus, ribosomes

☐ **D** cell wall, mitochondria, ribosomes

> Look at the mark allocation for each question – here there is one mark so you need to put a cross in **one** box.

(1 mark)

(b) Which of the following are found only in plant cells?

☐ **A** cell membrane, nucleus, chloroplast

☐ **B** cell membrane, vacuole, chloroplast

☑ **C** cell wall, chloroplast, vacuole

☐ **D** cytoplasm, chloroplast, vacuole

> Always answer multiple-choice questions, even if you don't actually know the answer.

(1 mark)

2 (a) Explain why muscle cells contain many mitochondria.

The carry out respiration To get energy ①
They need energy to contract

(2 marks)

(b) Explain why all plant cells contain mitochondria but only some contain chloroplasts.

> Chloroplasts need light to carry out photosynthesis. Use the function of a chloroplast to explain why you would not find them in certain cells, such as root cells.

*All plants have mitochondria because
They need it for releasing energy but only some
contain chloroplasts as they need light to carry out*
photosynthesis but Root cells don't get light

(2 marks)

3 Describe the difference between the functions of a cell membrane and a cell wall.

Guided

Cell membrane controls *what enters and leaves
while cell wall is there for structure*

(2 marks)

4 The main function of fat cells is to store fat. Pancreatic exocrine cells secrete pancreatic juice, which contains many different digestive enzymes.

> Do not be put off by 'pancreatic exocrine cells'. Remember that enzymes are proteins and consider where in the cell proteins are made.

Suggest an explanation for why pancreatic exocrine cells contain many more ribosomes than fat cells.

*proteins are Made in the ribosomes and the
fat cell only holds fat whereas the pancreatic
exocrine cell actually release things*

9/10

(2 marks)

Different kinds of cell

1   The genes in a bacterial cell are contained:

☐ **A**   on a circular chromosome only

☐ **B**   on plasmids only

☑ **C**   on plasmids and a circular chromosome

☐ **D**   ~~in the nucleus~~ | It cannot be D because bacteria do not have nuclei. | **(1 mark)**

2   The diagram shows a sperm cell and a bacterium. Note that the drawings are not to the same scale.

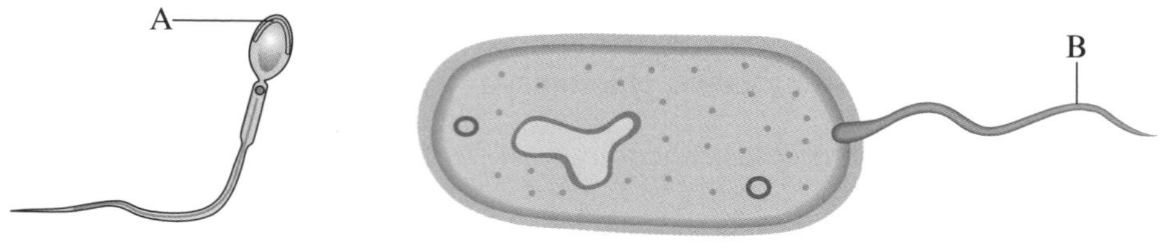

Sperm cell   Bacterium

(a)   Name the structures labelled A and B in the diagram:

A   acrosome

B   flagellum   **(2 marks)**

(b)   Describe the function of each structure.

A   The acrosome   penetrate   the   egg cells

B   The flagellum   is So the   Bacterium   can   move   **(2 marks)**

3   Breathing can expose us to dust, dirt and bacteria. Explain how cells in the lungs are specialised to protect us from these.

The ciliated epithelial cell   have   cillia which   helps catch the dust. Also the   alvea are   lined   with mucus So they catch   the dust and you   spit it out To get it out of the lungs   **(3 marks)**

4   The egg cell is much larger than the sperm cell. Give a reason to explain why.

The egg cell   contains   nutrients to help   the embryo grow.   **(2 marks)**

Microscopes and magnification

Guided

1 Scientists use two types of microscope to examine cells: light microscopes and electron microscopes. Describe how these types of microscope are different.

Light microscopes magnify...less........than electron microscopes.

The level of cell detail seen with an electron microscope is ...×..10.million.....

because ......they..use..electrons..to.view..enabled..... **(3 marks)**

2 The image shows an electron micrograph of part of a human liver cell.

   (a) Explain why this is a eukaryotic cell.

..It.is.a.eukaryotic.as.it...........
..Contains.a.nucleus.......................

....**(2 marks)**

mitochondrion

nucleus

2 µm

   (b) Estimate the size of the following parts of the cell:

     (i) the nucleus

.... **(2 marks)**

     (ii) the mitochondrion

.... **(2 marks)**

   (c) Explain why it would be possible to see the nucleus clearly using a light microscope, but the mitochondria would be unclear.

...A.nucleus..is.the..biggest..part.of..the.cell.........
..and.a.light.microscope.only.goes.too.×2000.however........
..mitochondrion.is.too.small.to.be.seen........................... **(3 marks)**

3 A scientist wants to study some bacteria that are 2.5 µm long. She can use either a light microscope (the one in the lab has a magnification of ×1000) or an electron microscope (the one in the lab next door has a magnification of ×100 000).

   (a) Calculate the size of the magnified image of the bacteria seen with each type of microscope.

     2.5 × 1000

       1×10^{18}

> **Maths skills** Remember that 1 µm = 1 × 10^{-6} m and do a reality check on your answer. The magnified image must be **bigger** than the bacteria and the image formed by the electron microscope must be **bigger** than that formed by the light microscope.

                                                     (3 marks)

   (b) Explain which microscope would be better for her to use.

...The.electron.microscope.would.be.better...
..to.use.as.study.the.nucleus.and.you...
..want.to.see.the.nucleus.def.....

> State which is better **and** give a reason.

                                                    (2 marks)

Dealing with numbers

Guided

1   Give the following units in order of increasing size:

   metre   micrometre   millimetre   nanometre   picometre

   picometre.. metre   **(1 mark)**

Guided

2   Convert the following quantities:

Quantity	Converted quantity
0.005 nanometres	5 picometres
250 milligrams	grams
250 milligrams	kilograms
2.5 metres	millimetres

   (4 marks)

Guided

3   For each of the following conversions, state whether it is true or false.

Conversion	True or false?
$0.000\,125\,mm = 0.125\,\mu m$	true
$150\,000\,mg = 0.015\,kg$	
$1\,kg = 10\,000\,000\,\mu g$	
$0.25\,mm = 2.5 \times 10^2\,\mu m$	

   (4 marks)

4   Calculate for each of the following the actual size of the structure in metres (m) **in standard form** to **two significant figures**.

> **Maths skills**   You have to remember whether to multiply or divide (check back on page 3 of the Revision Guide) as well as get the standard form right **and** round to 2 significant figures (one place of decimals in standard form).

   (a)   a ribosome that measured 30.9 mm in an electron micrograph (magnification = ×1 000 000)

   .... m   **(2 marks)**

   (b)   a mitochondrion that measured 163 mm in an electron micrograph (magnification = ×250 000)

   .... m   **(2 marks)**

   (c)   a nucleus that measured 7.8 mm in a light microscope (magnification = ×800)

   .... m   **(2 marks)**

Using a light microscope

1 (a) State the function of the following parts of a light microscope.

  (i) the mirror

  .... **(1 mark)**

  (ii) the stage with clips

  .... **(1 mark)**

  (iii) the coarse focusing wheel

  .... **(1 mark)**

  (b) Give the reasons for the following precautions when using a light microscope.

  (i) Never use the coarse focusing wheel with a high power objective.

  ....

  .... **(1 mark)**

  (ii) Never point the mirror directly at the Sun.

  ....

  .... **(1 mark)**

  (c) (i) State an alternative light source that might be safer than the Sun.

  .... **(1 mark)**

> **Guided**

  (ii) State **two** other precautions that you should take when using a light microscope.

  precaution 1   Always start with the lowest power objective under the eyepiece.

  precaution 2   ....

  .... **(2 marks)**

2 You are observing a slide under high power but cannot see the part you need.

Describe how you would bring the required part into view.

> **Practical skills** Think about why you cannot see what you need and then the steps you must follow to find it. Remember some of the precautions you have to take.

....

....

....

....

....

.... **(3 marks)**

Practical skills **Drawing labelled diagrams**

1   A student was given the slide below left and told to make a high power drawing to show cells in different stages of mitosis. His drawing is shown below right.

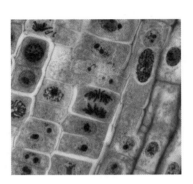   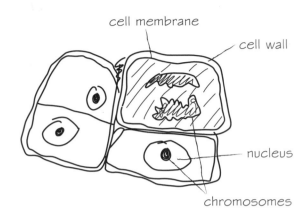

> **Guided**

(a)   Identify three faults with the student's drawing.

fault 1   The drawing is in pen rather than in ....

fault 2   ....

fault 3   ....   **(3 marks)**

(b)   Draw your own labelled diagram of the slide above.

> **Practical skills** Include outlines of all cells with more detail of cells showing different stages of mitosis. Try to show one of each stage.

(4 marks)

2   The student used a scale to measure the actual width of the field of view shown in the slide (above left) and found it was 0.113 mm. Calculate the magnification.

magnification = ...   **(3 marks)**

Enzymes

1 The enzyme invertase digests sucrose to glucose and fructose. Explain why invertase will not digest the sugar lactose.

> Guided

The shape of ....Substrab.......................... matches the shape of
.........active site.......................... so .....without the substrab..... cannot
combine with ......active site................

....

....

....
(2 marks)

2 The graph shows how the rate of an enzyme reaction changes with temperature.

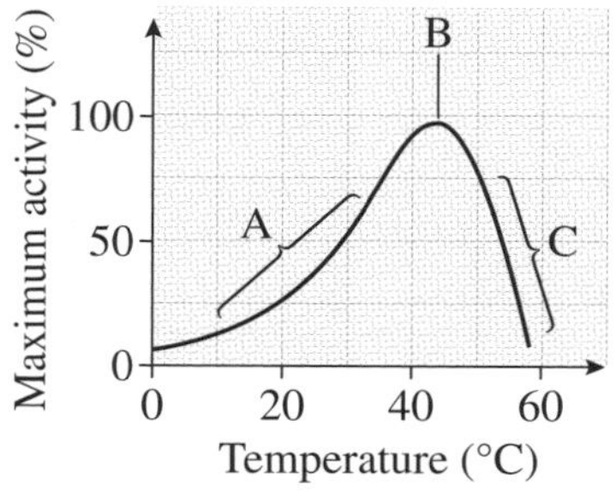

(a) Describe how the rate of reaction changes with temperature.

...When the rate of reaction is going...
...up and the reaction span...
...up than its at its optimum where...
...is fastest then goes to it slowest at...
(3 marks)

(b) Explain the effect of temperature on the rate of reaction in the following areas of the curve.

> Relate your explanation to the shape of the curve. Consider the effect of temperature on protein structure.

(i) region A ...It is increasing.......................
....
(2 marks)

(ii) point B ...fastest point.......................
....
(2 marks)

(iii) region C ...Slowest point.......................
....
(2 marks)

3 Pepsin and trypsin are proteases. Pepsin is produced in the stomach (pH 2), and trypsin is found in pancreatic juice (pH 8.6) released into the small intestine. Saliva and pancreatic juice both contain amylase. The graph shows the effect of pH on the activity of these enzymes.

Use this information to explain why proteins are digested in the stomach and small intestine, but starch is only digested in the mouth and small intestine.

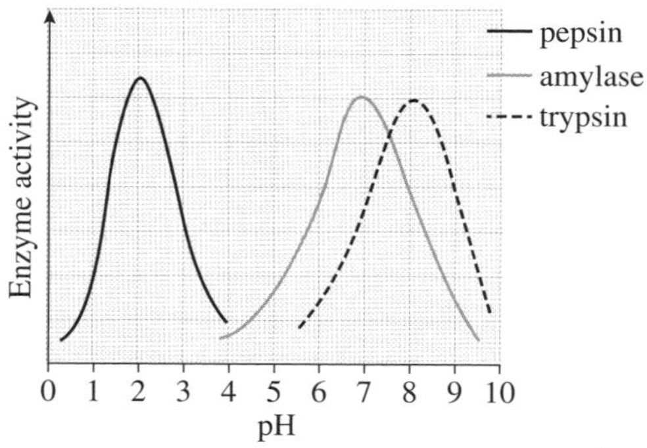

...Proteins are digested in the...
...Stomach as that is the optium...
...PH while Starch has its to
...take in the mouth...

....

....
(4 marks)

**Practical
skills** # pH and enzyme activity

1 A student carried out an experiment to
investigate the effect of pH on the activity
of the enzyme trypsin using pieces of
photographic film. Trypsin digests the protein
in the film and causes the film to turn clear.
Measuring the time it takes for the film to clear
allows you to calculate the rate of reaction. The
student used the apparatus shown.

This procedure was repeated using trypsin
solution at different pH values. The student's
results are shown in the table.

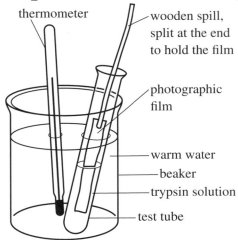

pH	2	4	6	8	10
Time (min)	> 10	7.5	3.6	1.2	8.3
Rate/min	O	0.13			

> Remember, rate = 1/time

Guided

(a) Complete the table by calculating the rate of reaction at each pH. **(2 marks)**

(b) Draw a suitable graph to show the effect of
pH on the rate of reaction.

> A suitable graph would be rate of reaction against
> pH. Make sure you include a title and label the axes!

(4 marks)

(c) State **two** ways in which the experiment could be improved.

improvement 1 ....

improvement 2 .... **(2 marks)**

The importance of enzymes

Guided

1 Complete the following table.

Enzyme	Digests	Product(s)
amylase	starch	
lipase		
protease		amino acids

(3 marks)

2 (a) Explain why different digestive enzymes are needed in the digestive system.

....

....

....

.... **(2 marks)**

(b) Explain the importance of enzymes as biological catalysts in building the molecules needed in cells and tissues.

....

....

....

.... **(2 marks)**

3 Biological washing powders contain enzymes that help to break down food stains on clothes.

(a) Eggs are rich in protein. Explain what type of enzyme is needed to remove egg stains from clothes.

....

....

....

.... **(2 marks)**

(b) Explain why biological washing powders work better below 40 °C.

> Think about what biological washing powders contain and what effect temperature might have.

....

....

....

....

.... **(3 marks)**

Getting in and out of cells

1 Define diffusion.

....

....

....

.... **(2 marks)**

Guided

2 Compare and contrast **diffusion** and **active transport**.

Both ....

....

Active transport requires ....

.... **(2 marks)**

Guided

3 (a) Explain what is meant by the term **osmosis**.

> Make sure you use the terms 'water', 'partially permeable membrane' and 'movement' in your answer.

Osmosis is the net movement of .... across a

....

from a low ....

to a high .... **(4 marks)**

(b) The blood in the lungs contains less oxygen and more carbon dioxide than the air. Explain why oxygen moves from the air to the blood and carbon dioxide moves from the blood to the air.

....

....

....

....

....

.... **(3 marks)**

(c) When starch is digested to glucose it is important that all the glucose is absorbed from the small intestine. Explain why this process requires energy.

....

....

....

.... **(2 marks)**

 **Osmosis in potatoes**

Guided

1   Describe how you would investigate osmosis in potatoes using potato pieces. You are provided with solutions of different sucrose concentrations. You should include at least **two** steps that you should use to ensure the accuracy of your results.

Cut pieces of potato, making sure ....

....

....

Remove from the solution, then ....

.... **(4 marks)**

2   The table shows the results of an experiment to investigate osmosis in potatoes.

Sucrose concentration (mol dm⁻³)	Initial mass (g)	Final mass (g)	Change in mass (g)	Percentage change (%)
0.0	19.15	21.60	2.45	12.8
0.1	18.30	19.25	0.95	
0.2	15.32	14.85	−0.47	
0.3	16.30	14.40	−1.90	
0.5	18.25	16.00	−2.25	
1.0	19.50	17.20	−2.30	

Guided

(a)   Complete the table by calculating the percentage change in mass. Use the space below for your working.

> You probably wouldn't have to do as many calculations as this in the exam, but it is good practice!

(2 marks)

(b)   (i)   Use the axes below to plot a graph of these results.

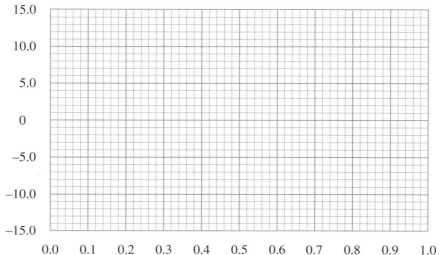

> Remember to label the axes, including the units used.

(3 marks)

(ii)   Use your graph to estimate the solute concentration of the potato cells.

solute concentration of potato cells = ... mol dm⁻³   **(1 mark)**

Extended response – Key concepts

The uptake of substances by yeast cells was studied using this method.

1. Add red dye to a suspension of yeast cells. Incubate the cells at 25 °C and observe any changes.

2. Repeat step 1 but incubate the cells at 5 °C.

3. Heat some yeast cells to 60 °C for 2 minutes, then cool them before repeating step 1.

The table shows the results.

Temperature of suspension	Appearance of yeast cells	Appearance of solution
25 °C	red after 30 minutes	colourless after 30 minutes
5 °C	colourless after 30 minutes	still red after 30 minutes
	red after 120 minutes	colourless after 120 minutes
heat-treated at 60 °C then cooled to 25 °C	colourless after 120 minutes	still red after 120 minutes

Explain what this experiment shows about the movement of coloured dye into the yeast cells. Describe other experiments you could do to confirm your conclusion.

You will be more successful in extended response questions if you plan your answer before you start writing. The question asks you to draw conclusions from the data and explain what is happening. Think about the appearance of the solution as well as the appearance of the cells.

Your answer should include the following:

• Describe ways in which the dye could enter the cells – explain the significance of the process being slower at 5 °C than at 25 °C.

• Explain the effect heating would have on the yeast cells. You need to use scientific language, not just 'It killed the cells'.

• Make sure that you include suggestions for further experiments, such as how you might use a microscope to look at what is happening.

Do not forget to use appropriate scientific terminology. Here are some of the words you should include in your answer:

active transport   enzymes   concentration gradient   denatured

....

....

....

....

....

....

....

....

.... **(6 marks)**

Continue your answer on your own paper. You should aim to write about half a side of A4.

Mitosis

1 (a) Read the following statements about mitosis. Which statement is correct? Tick **one** box.

☐ **A** A parent cell divides to produce two genetically different diploid daughter cells.

☑ **B** A parent cell divides to produce two genetically identical diploid daughter cells.

☐ **C** A parent cell divides to produce four genetically identical haploid daughter cells.

☐ **D** A parent cell divides to produce four genetically different haploid cells. **(1 mark)**

(b) List the stages of mitosis in the order they happen, starting with interphase.

...Prophase, Metaphase, Anaphase, Telophase... **(1 mark)**

2 A cell divides by mitosis every hour. State how many cells there will be after four hours.

Guided

To start with there is 1 cell; after 1 hour this divides into 2 cells. After 2 hours 4 cells. After 3 hours 8 cells. After 4 hours ....16 cells.......... **(1 mark)**

3 The photograph shows a slide of cells from an onion root tip at different stages in mitosis.

(a) Name the two stages of mitosis labelled A and B.

A ...Metaphase ✗ Anaphase...

B ...Interphase ✗ Metaphase... **(2 marks)**

(b) Give a reason for each answer in part (a).

A ...The Cell is Starting to split...

B ...The Cell is Starting To donh... **(2 marks)**

(c) In filamentous algae, telophase is not followed by cytokinesis. State what the result will be.

> Telophase ends when the nuclear membranes re-form. Think about what normally occurs next and what might be the result if it didn't happen.

...If it didn't happen Anaphase where the chromatids Separate and they are pulled to one Side of a Cell... **(2 marks)**

Cell growth and differentiation

1 (a) Give the name of a fertilised egg in animals.

.... ~~embryo~~ Zygote .... **(1 mark)**

(b) State the type of cell division that occurs after an egg is fertilised.

.... Mito mitosis ✓ .... **(1 mark)**

2 Plant cells divide by mitosis.

(a) State the name of the type of plant tissue where mitosis occurs rapidly.

> Remember that plants grow when cells divide and when cells elongate.

.... The tips of shoots and roots ✓ .... **(1 mark)**

(b) Describe how plant cells increase in size following mitosis.

~~Plant increase~~ in size outs step drawing when

growth n as plants aspects & when repair is need **(2 marks)**

3 (a) Complete the table to show whether the different specialised cells are animal or plant cells.

Sometimes the control that which when to sto given to sy wring

x Vacibles bok water

cansing the c ds to
e longe

Type of specialised cell	Animal or plant
sperm	animal
xylem	Plant
ciliated cell	animal
root hair cell	Plant
egg cell	animal

(3 marks)

(b) Give the name of one other type of specialised cell found in **plants** and one in **animals**.

plants ...~~root hair cell~~.........................

animals ...~~Sperm cell~~......................... **(2 marks)**

4 Growth in animals happens over a particular period of the animal's lifespan. Growth happens through cell division and when cells in the animal differentiate.

(a) Explain what is meant by the term **differentiate**.

...differentia........ become specialised to do a action proce

....

.... **(2 marks)**

(b) Explain why cell differentiation is important in animals.

Cell differntion creates specialised cells adapted to

carry out in particular functi.

....

.... **(2 marks)**

Growth and percentile charts

1 A midwife will measure the growth of a baby in different ways. The graph shows some percentile charts for the head circumference measurement for young children.

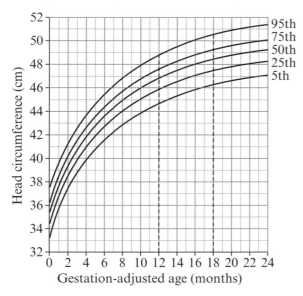

> Graphs like this sometimes look complicated – but remember that the curves are all labelled, so you can see what each one refers to.

> **Maths skills** Note that dashed guidelines have been put in to help you answer part (b). These help to show how you get the reading for both measurements from the graph – you can then subtract one number from the other to get the final answer.

(a) The median head circumference is described by the line where half the babies have a greater circumference, and half have the same or a smaller circumference. Which percentile curve shows the median rate of growth for babies?

☐ **A** 5th percentile   ☐ **B** 25th percentile   ☒ **C** 50th percentile   ☐ **D** 75th percentile

(1 mark)

(b) Use the graph to calculate the change in head circumference for a baby that lies on the 25th percentile curve between 12 and 18 months old. Show your working.

change in circumference ......1.5 6.......... cm   **(2 marks)**

2 Growth in seedlings can be investigated by measuring the mass of seedlings of different ages.

> Guided

(a) One seedling increased in mass from 12.75 g to 15.35 g over a period of 7 days. Calculate the percentage increase in mass for this seedling. Show your working.

15.35 – 12.75 = ...2.6........ g

$\left(\dfrac{...2.6...}{12.75}\right) \times 100 = ...%.x...... \%$

(2 marks)

(b) Describe **one** other way you could measure the growth of the seedlings.

...In ellarse in Size...

....

....

.... **(2 marks)**

Stem cells

> Guided

1 (a) In animals, stem cells are found in both adults and embryos.

Describe **two** ways in which adult and embryonic stem cells are different from each other.

> When you are asked to describe differences, remember that for each difference you have to say something about both the things you are comparing.

All the cells in an embryo are ....*stem cells*...., but in an adult, stem cells are

found only ...*differentiated tissue*....

Embryonic stem cells can differentiate into....*scales diploid*....

but adult stem cells can only differentiate into ...*damaged cells one type of cell*.... **(2 marks)**

(b) (i) Give the name of the tissue where plant stem cells are found.

....*Meristems*.... **(1 mark)**

(ii) Name **two** places in a plant where you would find stem cells.

....*roots*.... and ....*shoots* 2.... **(2 marks)**

2 (a) Describe **one** function of adult stem cells.

....*They divide to replace damaged cells*.... **(1 mark)**

(b) Describe **one** difference between an embryonic stem cell and a differentiated cell.

....*differentiated cells cannot divide but embryos can*.... **(1 mark)**

> To answer the following questions, think about what happens in a tissue transplant as well as what the different types of stem cell are capable of.

3 Parkinson's disease is caused by the death of some types of nerve cells in the brain.

(a) Describe how embryonic stem cells could be used to treat Parkinson's disease.

....*They can replace brain cells & can be stimulated to produce nerve cells*.... **(2 marks)**

(b) Another treatment method involves taking the patient's own cells (e.g. skin cells) and turning them into a type of stem cell called IPSCs. Give **one** advantage and **one** disadvantage of each method.

(i) embryonic cells

advantage: ....*easy to extract*....

disadvantage: ....*embryo destroyed*.... **(2 marks)**

(ii) IPSCs

advantage: ....*replacing faulty cells & dont destroy embryos*....

disadvantage: ....*Stem cells may not stop dividing causing cancer*.... **(2 marks)**

Neurones

1　State the function of each of the following.

> Don't say that neurones 'carry messages'; you have to be more specific and talk about electrical impulses.

(a) dendron ...Carries Impulses towards cell body............. **(1 mark)**

(b) axon ...Carries Impulses away from cell body............. **(1 mark)**

2　The diagram shows a sensory neurone.

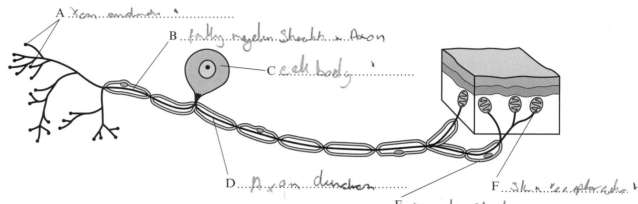

A ..Xon endings..

B ..fatty myelin Sheath x Axon

C ..cell body..

D ..Axon dunction..

E ..Myelin sheath..

F ..Skin receptor..

Label the parts A – F of the sensory neurone. Write your answers on the diagram. **(3 marks)**

3　(a)　Describe one way that the structure of a sensory neurone differs from the structure of a motor neurone.

Guided

The cell body of a sensory neurone is ...Is the middle of an axon..........

....

The cell body of a motor neurone is ...at the begining of the neuron............

.... **(2 marks)**

(b)　Explain how the structure of a motor neurone is related to its function.

...The Motor neurons carry impulses from the central nervous...

...system to effector organ and the axon carries the electral...

...Impulse.. **(3 marks)**

4　The table shows the speed at which nerve impulses are carried along two types of neurone.

Type of neurone	Speed of transmission (m/s)
myelinated	25
unmyelinated	3

(a)　Explain why the speed of transmission is different in the two types of neurone.

...There is an Insulation in the myelinated neurone so it...

...gets to across the fatty sheath where the myelinated cell are.. **(2 marks)**

(b)　In multiple sclerosis (MS), the myelin sheath surrounding motor neurones is destroyed. Explain what effect this would have on the movement of a person with MS.

...The electrical impulses cant cross but so it cant insulate the...

...neurone as It would be Slower.................................. **(2 marks)**

Responding to stimuli

1 The diagram shows a junction where neurone X meets neurone Y.

electrical impulse
axon of neurone X
gap between neurone X and neurone Y
neurone Y
electrical impulses to muscle

(a) State the name given to the junction between two neurones.

.........A.vev.\i................................... **(1 mark)**

(b) Explain which neurone (X or Y) on the diagram is a motor neurone.

....

.... **(2 marks)**

▷ **Guided** ▷ (c) Describe how neurones X and Y communicate.

When an electrical impulse reaches the end of neurone X it causes the release

of .... into the gap between the neurones. This substance ....

.... across the .... and causes neurone Y to ....

.... **(4 marks)**

2 The diagram shows a reflex arc.

(a) Describe the pathway taken by the nerve impulse in this reflex arc.

stimulus
sensory neurone
central nervous system
effector organ – muscle in the eyelid

....

....

....

....

.... **(4 marks)**

(b) What is the stimulus in this reflex arc? Give a reason for your answer.

| 'Give a reason' means you have to say something that supports your answer. |

....

.... **(2 marks)**

3 Explain the survival advantage of reflex responses.

....

....

....

....

.... **(3 marks)**

Extended response –
Cells and control

Fertilisation of a human egg cell produces a zygote, a single cell that eventually gives rise to every different type of cell in an adult human.

Describe the role of mitosis in the growth and development of a zygote into an adult human.

You will be more successful in extended response questions if you plan your answer before you start writing. Take care, because the question mentions fertilisation but it is really about growth and specialisation. Do not be tempted to talk about sexual reproduction – that is in the next topic.

Your answer should include the following:

• mitosis and cell division causing growth (from embryo to adult), and its importance in repair and replacement of cells

• cell differentiation to produce specialised cells

• the role of stem cells in the embryo as well as in the adult.

Do not forget to use appropriate scientific terminology. Here are some of the words you should include in your answer:

| cell cycle | replication | diploid | daughter cells | specialise | differentiate |

....

....

....

....

....

....

....

....

....

....

....

....

....

.... **(6 marks)**

Meiosis

1 Human gametes are haploid cells. During sexual reproduction, the gametes fuse to produce a zygote.

   (a) Describe what is meant by:

   (i) haploid

   .... **(1 mark)**

   (ii) gametes

   .... **(1 mark)**

   (b) State the name of the male sex cells and the female sex cells in humans.

   male ....

   female .... **(2 marks)**

2 A cell contains 20 chromosomes. It divides by meiosis.

   (a) State the number of chromosomes in each daughter cell.

   .... **(1 mark)**

   (b) Explain why the daughter cells are not genetically identical.

   ....

   .... **(2 marks)**

3 The diagram below shows a cell with two pairs of chromosomes undergoing meiosis.

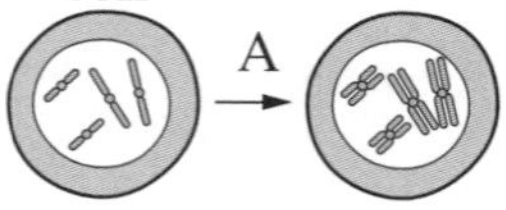

parent
cell

   (a) State the name of the process indicated by letter **A** in the diagram.

   .... **(1 mark)**

   (b) Complete the diagram above to show how daughter cells are formed.   **(3 marks)**

   > Use the drawing as a guide. Make sure you draw the chromosomes as they are shown, paying attention to the relative sizes.

4 Describe the importance of the two types of cell division, mitosis and meiosis.

   Guided

   Mitosis maintains the .... and produces cells that are

   .... to the parent cell. It is used for ....

   Meiosis creates .... that have .... the

   number of .... Fertilisation restores the

   .... **(5 marks)**

DNA

1 Our chromosomes contain genetic information. This information is held in our DNA.

(a) State the name used to describe all the DNA of an organism.

.... genome .... **(1 mark)**

(b) Describe the difference between chromosomes, genes and DNA.

> Guided

> This question is best answered by thinking of the definition of each of these terms.

A chromosome consists of a long molecule of .... genes .... which is ...
... a short piece of DNa that codes ...
... for a specyfic proteine .... **(3 marks)**

2 (a) What name is given to the shape of a DNA molecule?

.... Pauble helix .... **(1 mark)**

(b) The DNA molecule is made up of a series of bases.

(i) State the number of different bases present in DNA.

.... 2 Rpairs .... **(1 mark)**

(ii) Describe how the two strands of the DNA molecule are linked together.

.... By hydrogen bonds which ...
... hold it together .... **(1 mark)**

3 The diagram shows a section of DNA.

(a) DNA is a polymer. Give **one** piece of
evidence from the diagram that DNA is
a polymer.
.... It is bonded x ...
... it consists with ...
... nuchaloveru monomers ...... **(1 mark)**

A ... base ....

B. De Sugar ...

C. Phosphate ...

(b) Identify the components A, B and C of
the DNA structure. Write your answers
on the diagram.   **(3 marks)**

> You will not be expected to draw this structure from memory, but you may be expected to
> label the parts shown.

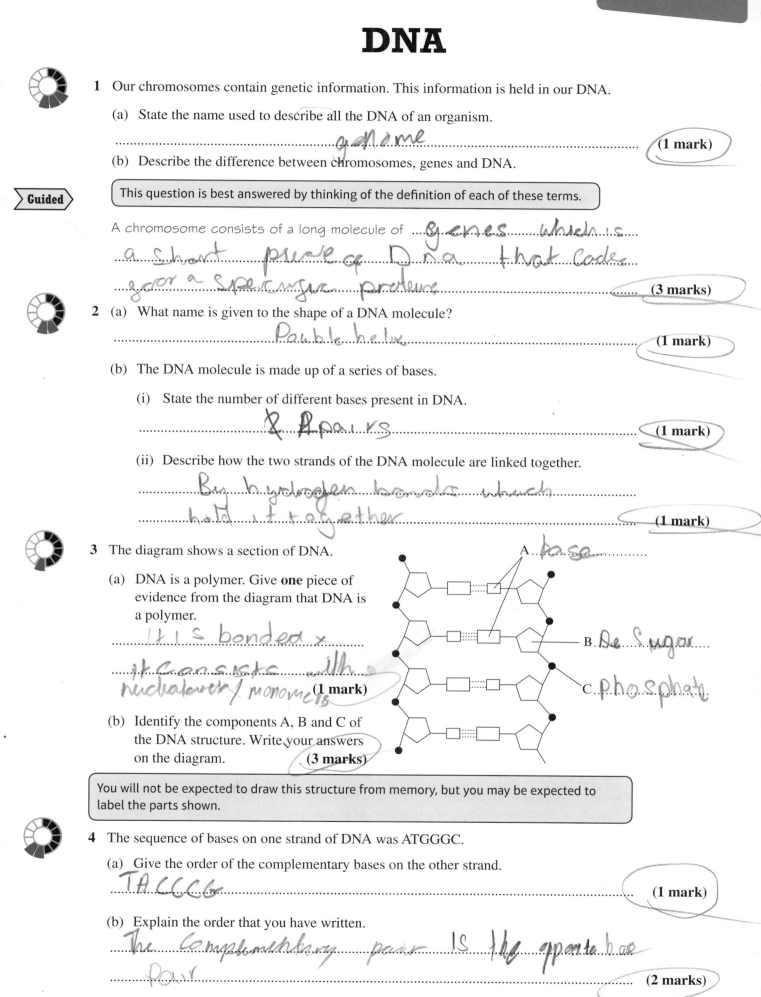

4 The sequence of bases on one strand of DNA was ATGGGC.

(a) Give the order of the complementary bases on the other strand.
... TACCCG .... **(1 mark)**

(b) Explain the order that you have written.
... The complementhbrry pair is the oppate bae ...
... Pair .... **(2 marks)**

21

Genetic terms

1 Eye colour in humans can be controlled by two alleles of the eye colour gene. One recessive allele (b) codes for blue and one dominant allele (B) codes for brown.

> You need to know what recessive, genotype, phenotype, homozygous and heterozygous mean.

(a) (i) State what is meant by alleles.

.... **(1 mark)**

   (ii) Using eye colour as an example, explain the difference between the terms **genotype** and **phenotype**.

....

....

....

.... **(2 marks)**

(b) State the following genotypes for eye colour:

homozygous blue: ....

homozygous brown: ....

heterozygous: .... **(3 marks)**

(c) A girl has blue eyes. Explain what her genotype must be.

....

....

....

.... **(2 marks)**

2 Mendel used the results from his experiments to devise his three laws of inheritance.

1. Each gamete receives only one factor for a characteristic.
2. The version of a factor that a gamete receives is random and does not depend on the other factors in the gamete.
3. Some versions of a factor are more powerful than others and always have an effect in the offspring.

Mendel did not know what these 'factors' actually were.

Explain how our understanding of genes and chromosomes has confirmed his laws.

There are two copies of each chromosome in body cells ....

....

....

....

.... **(4 marks)**

Monohybrid inheritance

1 Two plants both have the genotype Tt. The two plants are bred together.

The allele that makes the plants grow tall is represented by T and the allele that makes plants shorter is represented by t.

> **Maths skills** Percentage probabilities from Punnett squares will always be 0, 25%, 50%, 75% or 100%, depending on the number of squares with a particular genotype (0, 1, 2, 3 or 4 squares). In fractions, probabilities will always be 0, $\frac{1}{4}$, $\frac{1}{2}$, $\frac{3}{4}$, or 1.

(a) Complete the Punnett square to give the gametes of the parents and the genotypes of the offspring.

gametes of parent 1

gametes of parent 2

> Take great care to complete the square correctly and use the right letters.

(2 marks)

(b) State and explain the percentage of the offspring from this cross that will be short.

> **Guided**

25% of the offspring from this cross will be short. I know this because

....

.... **(3 marks)**

(c) Determine the probability of the offspring from this cross being tall. Express your answer as a fraction.

....

.... **(1 mark)**

2 Fur colour in mice can be represented by two alleles, G and g. Two parent mice were bred, and produced a total of 40 offspring. 50% of the offspring were white, which is the recessive characteristic and the rest were grey.

> **Guided**

Complete the genetic diagram to show this cross and show the genotypes of the parents.

> With this question it might be easier to start with what you know – the phenotypes of the offspring – and then work backwards.

Parent genotypes	....		....	
Gametes	....	....	....	....
Genotype of offspring	....	....	....	....
Phenotype of offspring	grey	white	grey	white

(4 marks)

Family pedigrees

1   Two healthy parents have a child who has sickle-cell anaemia, a condition caused by a recessive
    allele. Which **one** of the following is true?   **(1 mark)**

> Questions like this can be tricky! Some answers might be true in general, but not in this particular case.
> You need to pick the one that is true **and** applies to this example.

☐ **A** Both parents are homozygous for the sickle-cell allele.

☐ **B** One parent is homozygous for the sickle-cell allele and
          the other is homozygous for the normal allele.

☐ **C** Both parents are heterozygous for the sickle-cell allele.

☐ **D** One parent is heterozygous for the sickle-cell allele and
          the other is homozygous for the normal allele.

2   This family pedigree shows the
    inheritance of cystic fibrosis (CF).

    CF is a genetic condition in humans
    caused by a recessive allele.

    (a) State how many cystic fibrosis
         alleles an individual must
         inherit in order to show the
         symptoms of CF.

    ....
                                    (1 mark)

□ healthy male

○ healthy female

■ male with CF

● female with CF

    (b) State how many males in the
         family pedigree have a homozygous recessive genotype.

    .... **(1 mark)**

> Guided

    (c) State and explain the genotype of person 4. Use F for the normal allele and f for the recessive
         allele.

    Person 4 does not have cystic fibrosis. This means that they must have

    one .... allele from their father. But they must

    have inherited a .... allele from their mother.

    This means that their genotype is .... **(3 marks)**

    (d) Explain the evidence that cystic fibrosis is caused by a recessive allele.

> You have to look for patterns in problems like this. However, the fact that 1 and 2 had healthy children
> does not prove the allele is recessive. You need to look elsewhere.

    ....

    ....

    .... **(2 marks)**

Sex determination

1 (a) A baby girl is born. Explain which sex chromosome was in the sperm that fertilised the egg.

....

.... **(2 marks)**

Guided

(b) (i) Complete the Punnett square to show the sex chromosomes of both parents and all possible children.

This is a Punnett square but you could also use a genetic diagram to show how X and Y chromosomes combine.

Father

	X	
Mother		
X		

(2 marks)

(ii) State the sex of the child in the shaded box.

.... **(1 mark)**

2 (a) A couple who have a girl wish to have a second child. Explain the chance of the couple's second child being a boy.

....

....

....

....

....

.... **(3 marks)**

(b) Read this statement:

> If a couple have had children and they are all girls, then the next child is more likely to be a boy.

Discuss whether you think this statement is correct.

....

....

....

.... **(2 marks)**

Variation and mutation

Guided

1 Give the causes of differences between the following:

(a) the masses of students in a year 7 class

Students in a year 7 class will show differences in mass caused by .... variation as well as .... variation.   **(2 marks)**

(b) a pair of identical twins

Identical twins will only show differences caused by .... variation.   **(1 mark)**

2 Mr and Mrs Davies have six children. The table shows the heights of each of the six children when they reached adulthood.

Child	George	Arthur	Stanley	James	Josh	Peter
Adult height in cm	181	184	178	190	193	179

a) Calculate the mean height of the six Davies children. Show your working. Give your answer to 1 decimal place.

mean height = ... cm **(2 marks)**

(b) Mr Davies is 192 cm tall and Mrs Davies is 165 cm tall. Mr Davies wonders why his children show a range of different heights. Mrs Davies wonders why the mean height of their children is not the same as the mean of her height and her husband's height.
Suggest an explanation that will answer their questions.

> Don't forget to cover both genetic and at least one environmental factor. Make sure you use scientific language such as alleles and inheritance in your answer.

....

....

....

.... **(4 marks)**

3 (a) A plant growing in soil that was poor in nutrients had leaves that turned yellow after several weeks. When the plant flowered, the flowers were pink. Seeds were collected from the plant and grown the following year in soil that was rich in nutrients.
Explain why the leaves did not turn yellow but the flowers were still pink.

....

....

.... **(3 marks)**

(b) Cystic fibrosis is caused by a mutation that produces an inactive protein in the lung.
Explain how the cystic fibrosis mutation leads to production of an inactive protein.

....

....

.... **(3 marks)**

The Human Genome Project

1 (a) State what is meant by the human genome.

....

.... **(1 mark)**

> **Guided**

(b) State **two** advantages and **two** disadvantages of decoding the human genome.

advantage 1

A person at risk from a genetic condition will be ....

advantage 2

....

disadvantage 1

....

disadvantage 2

.... **(4 marks)**

2 The *BRCA1* mutation increases a woman's risk of developing breast cancer. Discuss the advantages and disadvantages to a woman of knowing that she has the *BRCA1* mutation.

> **Discuss** means you need to identify the issues being assessed by the question. You need to explore the different aspects of the issue. In this case, these are the advantages and disadvantages involved.

....

....

....

....

....

....

....

.... **(4 marks)**

Extended response – Genetics

The protein *p53* helps control cell division. There are many different mutations in the *p53* gene. People with a mutation of the *p53* gene are more likely to develop cancer. Discuss how the Human Genome Project can help identify those people who are at risk from cancer linked to mutant *p53*.

> You will be more successful in extended response questions if you plan your answer before you start writing.
>
> You should know that a mutation produces an allele that may change the phenotype – in this case, it is the risk of developing cancer.
>
> Some mutations can have a small effect on the phenotype, others can have a large effect.
>
> You need to link this to how the Human Genome Project helps to identify the different mutations and then link this to the risk of developing cancer.

....

....

....

....

....

....

....

....

....

....

....

....

.... **(6 marks)**

Evolution

1 Darwin proposed a series of stages in evolution, including genetic variation and environmental change.

(a) Describe what is meant by:

(i) genetic variation

.... **(1 mark)**

(ii) environmental change

.... changes In enviroment e.g new predator .... **(1 mark)**

(b) Explain why natural selection requires both genetic variation and environmental change.

..It requires genetic Variation as the genes need..
..to change and the enviromnt needs to chne..
..so the organisms occuring..
.... **(2 marks)**

2 Explain why, when an environment changes, some organisms within a species survive whereas others die.

> You should use scientific terms such as variation and survival in your answer.

..Individuals with Variations that are the less well adapted..
..to the enviroment will be less likely to Survive..
.... **(2 marks)**

3 When a new species is discovered, a scientist may take some of its DNA to analyse. Explain how this would help establish if this is a new species.

Guided

It will help .... Classify .... the new species and to find out

which other .... Species are related too .... **(2 marks)**

4 It is important to complete a course of antibiotics.

(a) Explain how stopping a course of antibiotics early can cause antibiotic resistance in bacteria.

> Darwin's theory was about natural selection and the survival of the fittest, so you should relate these to antibiotic resistance in bacteria.

..Bacteria that could be wiped out were W oder so they..
..evolved to be resistant against the antibiotic..
..to Survive.. **(4 marks)**

(b) Explain how this provides evidence for Darwin's theory of evolution.

..This prove that more individuals will have their..
..archangeos Variens In the next generation..
.... **(3 marks)**

Human evolution

1 Apart from the differences in body hair, using the diagrams of Ardi and Lucy, state three differences between them.

1 Ardi was 13cm taller

2 Ardis brain size was smaller

3 Ardi could climb trees but lucy can't

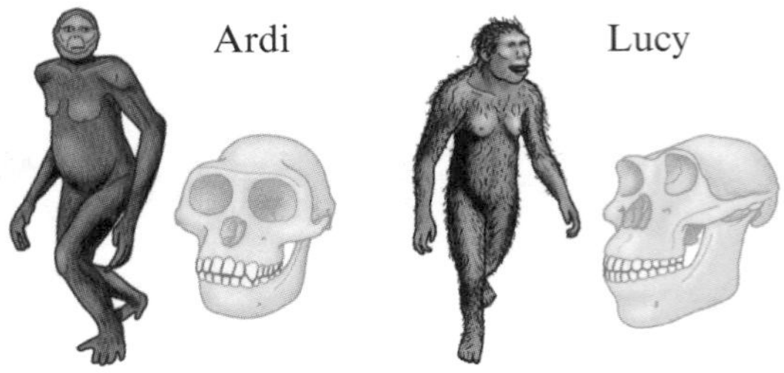

Ardi   Lucy

(3 marks)

2 Some evidence for human evolution has come from the fossil record of the skull. The table below shows some of this evidence.

> You do not need to remember details such as brain sizes but you do need to remember the names and the general trends.

Name of species	Year before present when species first appeared (millions of years ago)	Brain volume (cm³)
Ardipithecus ramidus (Ardi)	4.4	350
Australopithecus afarensis (Lucy)	3.2	400
Homo habilis	2.4	550
Homo erectus	1.8	850

(a) Describe the relationship between when each species first appeared and their brain volume.

The older the age the smaller the brain volume

(2 marks)

(b) The first stone tools are dated from about 2.4 million years ago. Using the table, deduce what may have enabled the use of stone tools.

> **Guided**

An increase in brain size would make the stone tools as the stone tools are much complicated **(2 marks)**

3 The diagram shows two images of stone tools.

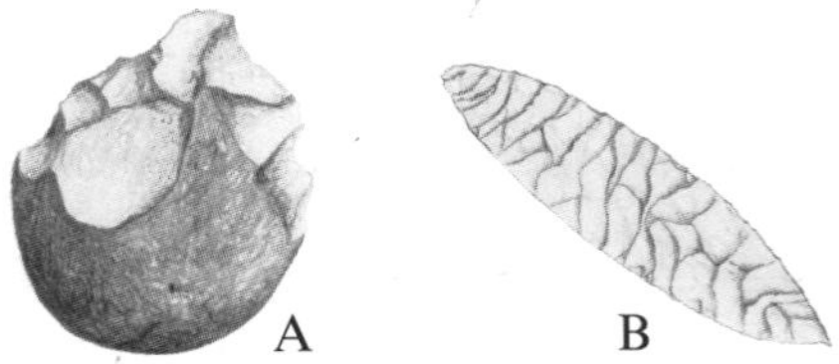

(a) Explain how scientists work out the ages of stone tools.

They measure the amount of radiation in all the samples of rock

(2 marks)

A   B

(b) Using the diagram, explain how stone tool A was held. Give reasons for your answer.

Stone A was grabbed and used to break very you can see

(3 marks)

Classification

1 Describe the similarities between a human arm and a bat's wing that suggest humans and bats share a common ancestor.

    Both have bones and can both move freely

    .... **(2 marks)**

2 Give **two** reasons why animals and plants are placed in separate kingdoms.

    Plants .... while animals ....

    Plant cells have a cell wall

    but animal cells dont .... **(2 marks)**

Guided

3 The table shows how some organisms are classified.

Classification group	Humans	Wolf	Panther
kingdom	Animalia	Animalia	Animalia
phylum	Chordata	Chordata	Chordata
class	Mammalia	Mammalia	Mammalia
order	Primate	Carnivora	Carnivora
family	Hominidae	Canidae	Felidae
genus	Homo	Canis	Panthera
species	Sapiens	Lupus	Pardus
binomial name	*Homo sapiens*	*Canis lupus*	*Panthera pardus*

Explain which two organisms in the table are most closely related.

    Wolf and panther as they go all the way to order wheras humens only goes to class

    .... **(2 marks)**

4 Carl Woese proposed that all organisms should be divided into three domains. Complete the table to give the missing information.

> Follow the example of the table and comment on the nucleus and genes.

Domain	Distinguishing characteristic of the domain
Archaea	mainly bacteria that live in warm or salty conditions
Eubacteria	
	cells with a nucleus, unused sections in genes

(3 marks)

Selective breeding

1 (a) Describe what is meant by selective breeding.

....

.... **(2 marks)**

(b) Explain how pig breeders could use selective breeding to produce lean pigs with less body fat.

> The principles of selective breeding are the same, even if you aren't familiar with this example.

....

....

....

....

.... **(3 marks)**

2 Food production can be increased by conventional plant breeding programmes.

(a) Explain **three** different characteristics that could be selected for in a crop suitable for use in any country.

....

....

....

....

.... **(3 marks)**

(b) Explain **two** other characteristics that might be selected for in a crop to be grown in a hot, dry part of Africa.

....

....

.... **(2 marks)**

(c) Give a reason why wheat in the United Kingdom has been selected to have a short stem length.

....

.... **(1 mark)**

3 Give **three** risks of selective breeding.

1 Alleles that might be useful in the future ....

2 ....

....

3 ....

.... **(3 marks)**

Genetic engineering

Guided

1   Scientists have produced mice that glow green in blue light. These 'glow mice' contain a gene naturally found in jellyfish. Explain why a glow mouse is described as a genetically modified organism.

Mice do not normally glow, but glow mice have a ....

....

.... **(2 marks)**

2   Scientists may genetically modify crop plants.

(a)   Name one crop plant that has been genetically modified.

.... **(1 mark)**

(b)   Describe two ways in which genetically modified crops benefit humans.

....

....

.... **(2 marks)**

(c)   Genetic modification can make crop plants resistant to insects by introducing certain bacterial genes.

Explain one disadvantage of doing this.

> Think about how these genetically modified plants may affect other living organisms in their environment.

....

....

....

....

.... **(2 marks)**

3   Human insulin can be produced by genetically modified bacteria. Discuss the advantages and disadvantages of this process.

> When asked to discuss, you need to give both sides of an argument. Here you should give at least one advantage and one disadvantage.

....

....

....

....

....

.... **(4 marks)**

Stages in genetic engineering

1 State the meaning of the following terms.

(a) plasmid

.... **(1 mark)**

(b) vector

.... **(1 mark)**

(c) sticky ends

.... **(1 mark)**

2 People with diabetes rely upon insulin. Human insulin can be produced by genetically modified bacteria, through genetic engineering. Explain the role of the following in the production of human insulin:

> Guided

(a) a human gene

The human gene needed is ....

It is needed because .... **(2 marks)**

(b) enzymes

....

.... **(2 marks)**

(c) bacteria

....

.... **(2 marks)**

3 (a) Explain why the same restriction enzyme is used to cut DNA from a human cell and to cut bacterial plasmids open.

> Remember that restriction enzymes produce DNA fragments with 'sticky ends'.

....

....

....

....

....

.... **(3 marks)**

(b) Describe the role of DNA ligase in genetic engineering.

....

....

....

.... **(2 marks)**

Extended response – Genetic modification

Compare and contrast the processes of evolution and selective breeding.

> You will be more successful in extended response questions if you plan your answer before you start writing.
>
> In this example, you need to say how these two processes are similar and how they are different.

....

....

....

....

....

....

....

....

....

....

....

....

....

.... **(6 marks)**

Health and disease

1 According to the World Health Organization (WHO), good health is a state of 'complete physical, social and mental well-being'. State what is meant by the following terms.

(a) physical well-being

...Being Well and healthy... **(1 mark)**

(b) mental well-being

...how you feel about yourself... **(1 mark)**

(c) social well-being

...how you get on with other people... **(1 mark)**

2 (a) Complete the table by putting a tick in the appropriate box to show whether the disease is communicable or non-communicable.

> Guided

Disease	Communicable	Non-communicable
influenza ('flu')	✔	
lung cancer		✓
coronary heart disease		✓
tuberculosis	✓	
Chlamydia (a type of STI)	✓	

(3 marks)

(b) Explain the difference between communicable and non-communicable diseases.

...A comicable disease can be passed on from person to person while a non-commicable disease is not passed from person to person **(2 marks)**

3 HIV is a virus that can infect humans. HIV makes it easier for other pathogens to infect the human body. Suggest an explanation for how HIV does this.

> Think about what type of cells are infected by the HIV virus.

...Hiv damages your cell while blood cells Making your Immune System Weaker **(2 marks)**

4 (a) Explain how viruses cause disease.

...Vixuses Take over your cells and r...

....

.... **(3 marks)**

(b) Describe **two** ways in which bacteria make us feel ill.

...bacteria Can make your cells do what there not Supposed too - or could kill the cells

.... **(2 marks)**

Common infections

1   The table shows the percentage of 15 to 49 year olds with HIV in some African countries.

	% of 15 to 49 year olds with HIV in some African countries			
African country	**2006**	**2007**	**2008**	**2009**
Namibia	15.0	14.3	13.7	13.1
South Africa	18.1	18.0	17.9	17.8
Zambia	13.8	13.7	13.6	13.5
Zimbabwe	17.2	16.1	15.1	14.3

(a)   Identify the country with the largest decrease in the percentage of HIV between 2008 and 2009. Show your working

> **Maths skills** First work out what the decrease was for each country. For example Zambia went from 13.6% to 13.5%. If you are not sure – use your calculator!.

country with largest decrease ....Zimbabwe.............. **(2 marks)**

(b)   The data for each African country follows the same overall trend. Use the data in the table to describe this trend.

......The percentage of each country comes on......
......decreasing...................................... **(2 marks)**

2   (a)   What kind of pathogen causes *Chalara* ash dieback?

> Guided

   ☐ **A** ~~a virus~~   ☐ **C** a protist

   ☒ **B** a bacterium   ☒ **D** a fungus   **(1 mark)**

(b)   Describe the effects of the pathogen on the trees.

...... It kills the trees and damages them......
.... **(2 marks)**

3   The table shows several diseases, the type of pathogen that causes them and the symptoms (signs of infection). Complete the table by filling in the gaps.

Disease	Type of pathogen	Signs of infection
cholera	bacterium	watery faeces
	bacterium	persistent cough – may cough up blood-speckled mucus
malaria	Virus	fart flu-like illness
HIV	bacterium	mild flu-like symptoms at first

(3 marks)

4   *Helicobacter* is a pathogen that causes stomach ulcers.

(a)   State the type of pathogen involved.

.... bacterium .... **(1 mark)**

(b)   Describe the symptoms it causes in infected people.

...... Inflamation. Bleeding in stomach...... **(2 marks)**

How pathogens spread

1 Which of these statements about malaria is correct?

☐ **A** Malaria is caused by a mosquito that invades liver cells.

☐ **B** The malaria pathogen is a mosquito.

☑ **C** The malaria pathogen is a protist that is spread by a vector, the mosquito.

☐ **D** The malaria pathogen is a mosquito that is spread by a vector, the protist.   **(1 mark)**

2 Complete the table to show ways in which the spread of certain pathogens can be reduced.

Disease	Pathogen	Ways to reduce or prevent its spread
Ebola haemorrhagic fever	Virus	keep infected people isolated; wear full protective clothing while working with infected people or dead bodies
tuberculosis	bacterium	Ventilate buildings so the tb bacteria drops dont fall.

(2 marks)

3 Cholera is a disease that can spread rapidly in disaster areas when drinking water supplies are damaged. Explain **one** way that its spread could be reduced or prevented.

If you boil the infected water it would work as the heat kills the bacteria

(2 marks)

4 (a) Explain why bacterial diseases such as cholera are less common in developed countries.

> Think about how these diseases are spread and how developed countries are able to control them.

the developed countries have better medical treatment so they can control it easier

(2 marks)

> Guided

(b) Explain why, during the 2014–15 Ebola outbreak, health workers wore full body protection when handling dead bodies.

To prevent being infected by ebola because it is a virus because Ebola virus is present in Infected people because it is easy to spread

(2 marks)

STIs

1   State what is meant by an STI.

....................Sexually Transmitted infection............................ **(1 mark)**

2   Which of these statements about *Chlamydia* is correct?

☐ **A**   *Chlamydia* is an STI caused by a virus.

☐ **B**   A person infected with *Chlamydia* may not realise they are infected.

☑ **C**   The number of new cases of *Chlamydia* diagnosed each year is falling.

☐ **D**   *Chlamydia* cannot be passed from mother to baby during birth.   **(1 mark)**

3   Complete the table.

> **Guided**

Mechanism of transmission	Precautions to reduce or prevent STI
unprotected sex with an infected partner	using condoms during sexual intercourse
Sharing of needles	supplying intravenous drug abusers with sterile needles
infection from blood products	use gloves when handilling

(3 marks)

4   (a)   Explain how screening for STIs can help to reduce transmission.

.........Screeninge for STIs can reduce.........
.........transmission as they can make it not go to hubdy **(2 marks)**

(b)   Many STIs can be treated with antibiotics. Explain why HIV cannot be treated with antibiotics.

> You will need information about treating infections to be able to answer this question. Review page 43 of the Revision Guide if you haven't already covered this.

.........HIV isnt a bacterium but a virus and.........
.........antibiotics only kill bacteria............... **(2 marks)**

Human defences

1 (a) Describe the role of the skin in protecting the body from infection.

It is to thick so viruses can't get through

(1 mark)

(b) Describe one chemical defence against infection from what we eat or drink.

Ly Zames

(1 mark)

(c) (i) Name an enzyme, found in tears, that protects against infection.

Lysase

(1 mark)

(ii) Describe how the enzyme named in part (i) protects the eyes against infection.

You need to name the enzyme and say what it does.

It kills bacteria

(2 marks)

2 The diagram shows a section of epithelium in a human bronchiole, one of the tubes in the lung.

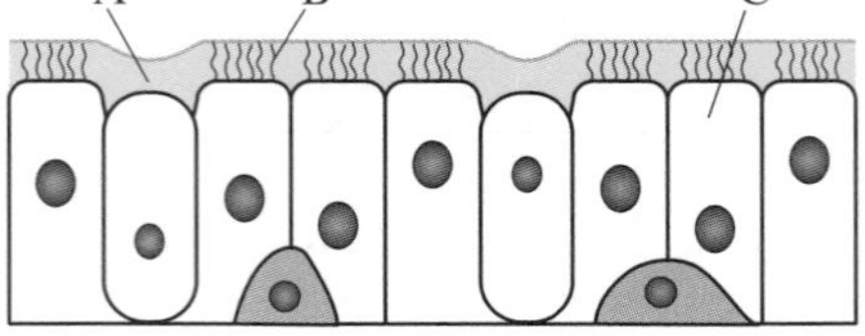

(a) (i) State the name of the substance labelled A.

epithil cells cillia

(1 mark)

(ii) Describe the role of substance A in protecting the lungs from infection.

Stops patho gens entering lung

(1 mark)

(b) (i) State the name of the structure labelled B.

Mucas layer

(1 mark)

Guided (ii) Describe the part played by the type of cell labelled C in protecting the lungs from infection.

The ...Mucur... on the surface of these cells move in a wave-like motion
because the pathogens are destroyed by the c
acid

(3 marks)

(c) Chemicals in cigarette smoke can paralyse the structures labelled B.

Explain why this increases the risk of smokers suffering from lung infections compared with non-smokers.

They have less of layer of protection meaning
More Weakness

(2 marks)

The immune system

1 Name the type of blood cell that produces antibodies.

...White blood cell... **(1 mark)**

2 Describe how lymphocytes help protect the body by attacking pathogens.

> **Guided**

Pathogens have substances called .Antiges.... on their surface. White blood

cells called .Platelets. are activated if they have .Subras that fit these

substances. These cells then ....

They produce large amounts of antibodies that .... **(5 marks)**

3 The graph shows the concentration of antibodies in the blood of a young girl. The lines labelled A show the concentration of antibodies effective against the measles virus. The line labelled B shows the concentration of antibodies effective against the chickenpox virus.

> There is a lot to think about in this question so take it one step at a time.

(a) At the time shown by arrow 1, there was an outbreak of measles. The girl was exposed to the measles virus for the first time in her life. Explain the shape of line A in the five weeks after arrow 1.

....

....

....

....

.... **(4 marks)**

(b) Five months later (shown by arrow 2) there was an outbreak of measles and chickenpox. The girl was exposed to both viruses.
Explain the shape of line A in the five weeks after arrow 2.

....

.... **(3 marks)**

(c) Use lines A and B to help you answer these questions.

(i) State whether the girl had been exposed to the chickenpox virus in the past. Explain your answer.

....

.... **(2 marks)**

(ii) In the second outbreak of measles, the girl showed no symptom of measles. Explain why.

....

.... **(2 marks)**

Immunisation

1 Children are immunised against many childhood infections.

(a) Explain what is meant by immunisation.

When you give a person a Vaccine to prevent them from being ill from a disease **(2 marks)**

(b) State what is meant by a vaccine.

A vaccine contains antigens from a pathogen, often in the form *of a dead or weakend Version*

(2 marks)

(c) Explain how a vaccine prevents a person from becoming ill from a disease, if they are exposed to the disease months or years after the vaccination.

Their body is used to the disease so their blood cells are already used to the diseas

(3 marks)

2 In 1998, a group of doctors suggested there was a connection between the MMR (measles, mumps and rubella) vaccine and autism. This made some parents afraid of having their babies vaccinated. The graph shows how the percentage of babies in the UK who were given the MMR vaccine changed over the following years.

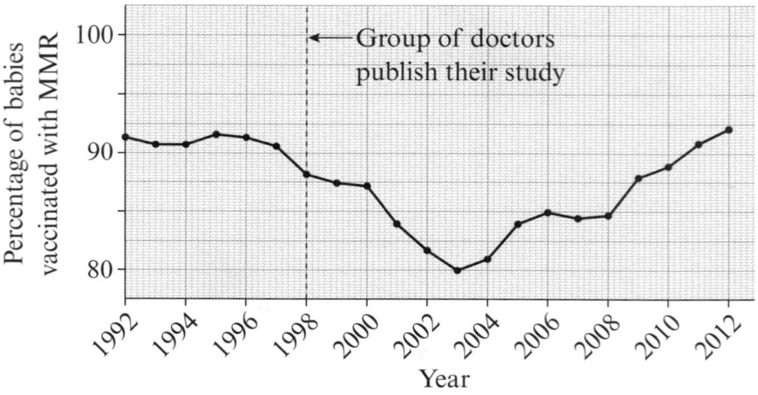

(a) State which year had the lowest rate of vaccination.

2003

(1 mark)

(b) Explain what you would expect to happen to the number of children suffering from measles in the period 1998–2004.

It would decrease

(2 marks)

Treating infections

1 (a) Which of the following statements is correct?

☐ **A** An antibiotic is produced in the body to fight infection.

☐ **B** Some antibiotics are becoming resistant to bacteria.

☐ **C** Antibiotics are medicines that kill or slow down growth of bacteria in the body.

☐ **D** Antibodies are medicines that kill or slow down growth of bacteria in the body.   **(1 mark)**

(b) Explain why antibiotics can be used to treat bacterial infections in people.

....

....

....   **(2 marks)**

2 Colds are caused by viruses. A man has a very bad cold. He asks a pharmacist if an antibiotic such as penicillin would help to cure his cold.

> Guided

State, with a reason, whether the pharmacist would advise the man to take penicillin.

The pharmacist's advice would be ....

The man's cold is due to a virus, so the penicillin ....

....   **(2 marks)**

3 Sinusitis causes a stuffy nose. Some patients with sinusitis were divided into two groups. One group was treated for 14 days with antibiotics while the other group did not receive antibiotics. Each day they were asked if they still had symptoms. The results are shown in the graph.

(a) State what you can deduce about the cause of sinusitis from the data.

> You are only asked for a deduction, not an explanation – although you might need to think about the answer to part (b) before you make your deduction!

....   **(1 mark)**

(b) Discuss whether the data supports the use of antibiotics to treat sinusitis.

> Be sure to refer to data in the graph when answering this question.

....

....

....

....   **(2 marks)**

New medicines

1 Development of a new medicine involves a series of stages. A new medicine can only move to the next stage if it has been successful in the previous stage.

Guided

(a) Complete the table to show the correct order of stages of developing a new drug.

Stage	Order
Testing in a small number of healthy people	
Discovery of possible new medicine	1
Given widely by doctors to treat patients	
Testing in cells or tissues in the lab	
Testing in a large number of people with the disease the medicine will treat	

(3 marks)

(b) (i) Describe **two** stages of preclinical testing in the development of a new medicine.

....

.... **(2 marks)**

(ii) Describe how development of a new medicine ensures that there are no dangerous side effects in humans.

.... **(1 mark)**

(c) Describe the function of a large clinical trial in developing a new drug.

> For three marks you will have to describe all of the functions; pay attention to the word 'large'.

....

....

.... **(3 marks)**

2 Scientists trialled a new medicine that was developed to lower blood pressure. They took 1000 people with normal blood pressure (group A) and 1000 people with high blood pressure (group B). Each group was divided in half; half the volunteers were given the new medicine and the other half were given a placebo (dummy medicine). At the end of the trial, the scientists measured the number of volunteers in each group who had high blood pressure.

The results are shown in the bar chart.

(a) Explain why it is important for medicine trials to use large numbers of volunteers.

....

.... **(2 marks)**

(b) Use information from the bar chart to evaluate the effectiveness of this medicine.

....

.... **(2 marks)**

Non-communicable diseases

1 Explain how an infectious disease is different to a non-communicable disease.

An infectious disease is caused by a .... and is passed from ....

.... A non-communicable disease

is not passed .... **(3 marks)**

2 State **three** factors that can affect a person's risk of developing a non-communicable disease.

1 ....

2 ....

3 .... **(3 marks)**

3 The two graphs show the prevalence of coronary heart disease (CHD) in men and women from different ethnic groups in the West Midlands. Prevalence means the percentage of people in that ethnic group who are diagnosed with the disease.

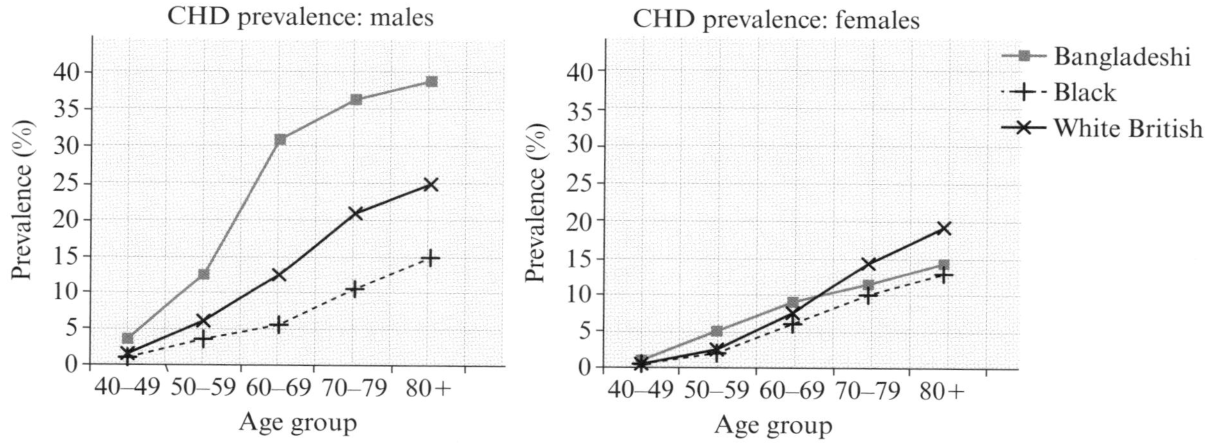

(a) State the group with the:

   (i) highest incidence of CHD .... **(1 mark)**

   (ii) lowest incidence of CHD... **(1 mark)**

(b) Discuss the effect of age, sex and ethnic group on the risk of developing CHD. Use the information in the graphs in your answer.

> **Discuss** means you need to identify the issues being assessed by the question. You need to explore the different aspects of the issue. In this case, how the incidence of CHD varies with age, sex and ethnic group.

> Make sure you cover all three factors (age, sex and ethnic group) as well as using data from the graph to support your conclusions.

....

....

....

....

....

.... **(4 marks)**

Alcohol and smoking

1 (a) Explain how alcohol (ethanol) causes liver disease.

....

....

....

....

....

.... **(3 marks)**

(b) State why alcohol-related liver disease is described as a lifestyle disease.

.... **(1 mark)**

2 Babies whose mothers smoked while pregnant have low birth weights. Explain why.

....

....

....

.... **(2 marks)**

3 (a) State **two** diseases caused by substances in cigarette smoke.

> The question asks you to state two diseases. Remember that heart attack or stroke are not diseases, they are the result of disease.

....

....

....

.... **(2 marks)**

⟩Guided⟩   (b) A stroke is caused by cardiovascular disease in the brain. Explain how smoking can lead to a stroke.

Substances in cigarette smoke cause blood vessels to ....

....

....

....

....

....

.... **(3 marks)**

Malnutrition and obesity

1 The graph shows the percentage of different age groups with anaemia in a population in the USA during the 1990s.

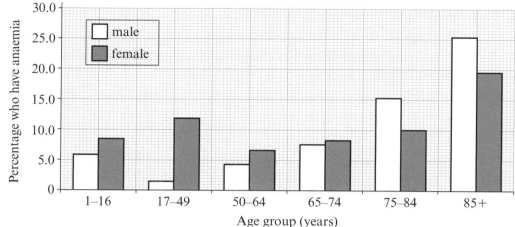

(a) Anaemia is a deficiency disease. State what is meant by deficiency disease.

.... **(1 mark)**

(b) Describe how the incidence of anaemia changes with age in males and females.

> Be sure to describe the trends in both males and females. Also, you are asked to describe – so do not try to explain!

....

....

....

.... **(4 marks)**

2 The table shows the height and mass of three people.

Subject	Mass (kg)	Height (m)	BMI
person A	80	1.80	24.7
person B	90	1.65	
person C	95	2.00	

(a) Complete the table by calculating the BMI for each person. **(2 marks)**

(b) Identify the person who is obese.

.... **(1 mark)**

3 Explain how measuring waist : hip ratio is better than BMI when predicting risk of cardiovascular disease.

....

....

....

.... **(3 marks)**

Cardiovascular disease

1 (a) State **two** ways in which cardiovascular disease may be treated.

....

.... **(2 marks)**

(b) State **two** pieces of advice a doctor might give to a patient with high blood pressure to help them to make lifestyle changes.

....

....

.... **(2 marks)**

(c) Explain why it is more important to prevent cardiovascular disease than to treat it.

....

....

.... **(2 marks)**

> **Guided**

2 The table summarises some of the benefits and drawbacks of the different types of treatment for cardiovascular disease.

Type of treatment	Benefits	Drawbacks
lifestyle changes	no side effects	may take time to work
medication	easier to do than change lifestyle	can have side effects
surgery	once recovered, there are no side effects	
		risk of infection after surgery

Complete the table with benefits and risks of the different types of treatment. **(3 marks)**

> **Guided**

3 Angina is chest pain caused by narrowing of the coronary arteries. This can be treated using a stent. A stent is a wire frame that is inserted into the narrowed part of the artery. Angina can also be treated using heart bypass surgery. This is where the narrowed artery is bypassed using a section of artery or vein.

> Remember that the coronary arteries are in the heart and supply heart muscle. Think about the consequences if they become blocked.

Evaluate the use of surgery to treat angina.

Surgery can help prevent .... but costs more than inserting a

.... and surgery ....

However, ...

....

.... **(4 marks)**

Extended response – Health and disease

Compare and contrast the causes and treatment of communicable and non-communicable diseases.

> You will be more successful in extended response questions if you plan your answer before you start writing.
>
> Make sure you compare and contrast both communicable and non-communicable diseases. This means you need to describe the similarities and differences between the causes and the treatments; try to link these together.

....

....

....

....

....

....

....

....

....

....

....

....

....

.... **(6 marks)**

Photosynthesis

1   Explain why it is that food chains start with plants or algae.

> Think about what a food chain represents. You will need to use terms such as producer and biomass in your answer.

....

....

.... **(3 marks)**

2   (a)   Complete the equation to show the reactants and products of photosynthesis.

Guided

      .... + water → .... + ...   **(2 marks)**

  (b)   Explain why photosynthesis is an endothermic reaction.

....

.... **(2 marks)**

3   A student knew that the products of photosynthesis are converted into starch in leaves. She also knew that iodine solution can be used to test for the presence of starch, producing a blue-black colour. She devised the following experiments to investigate photosynthesis.

Experiment 1

- A plant was kept in the dark for 48 hours to remove all starch from the leaves.
- Some of the leaves were covered in foil.
- The plant was then placed on a sunny windowsill all day.
- At the end of the day two leaves were tested for starch.
- One leaf had been covered in foil and did not produce a blue-black colour when tested with iodine.
- The other leaf had not been covered in foil and produced a blue-black colour when tested with iodine.

Experiment 2

- A plant with variegated (partially green, partially white) leaves was kept in the dark for 48 hours to remove all starch from the leaves.
- The plant was placed on a sunny windowsill all day.
- At the end of the day one leaf was tested for starch.
- Only the green parts of the leaf gave a positive test for starch.

  (a)   Explain what you can conclude about the requirements for photosynthesis from Experiment 1.

....

.... **(2 marks)**

  (b)   Explain what you can conclude about the requirements for photosynthesis from Experiment 2.

....

.... **(2 marks)**

Limiting factors

1 (a) Name **one** factor other than carbon dioxide concentration and light intensity that limits the rate of photosynthesis.

.... **(1 mark)**

(b) Describe how you could measure the rate of photosynthesis using algal balls.

....

....

.... **(3 marks)**

2 The graph shows how the rate of photosynthesis changes with light intensity. The data shows the rate at three different concentrations of carbon dioxide.

(a) Describe how increasing the concentration of carbon dioxide changes the rate of photosynthesis.

.... **(1 mark)**

(b) Commercial growers often increase the concentration of carbon dioxide in their greenhouses.

Explain how this will increase the yield of crops grown in the greenhouse.

....

.... **(2 marks)**

> **Guided**

(c) Explain how the rate of photosynthesis could be increased further.

You could increase the .... as this

would make photosynthesis happen.. **(2 marks)**

3 A farmer notices that when he changes the temperature of his greenhouse from 15 °C to 25 °C the plants grow more quickly. The plants grow at the same speed at 35 °C, but do not grow at all at 45 °C. The farmer knows that photosynthesis uses enzymes. Explain why the growth of the plants changes in this way.

> If you are asked to explain something make sure you do explain rather than just describe.

....

....

....

....

....

.... **(4 marks)**

Light intensity

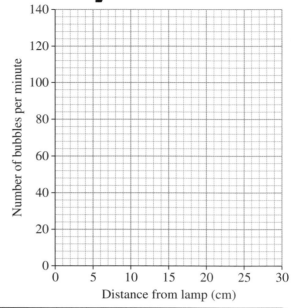

1   Some students wanted to investigate how the rate of photosynthesis in pond weed changed with light intensity. They did this by putting a lamp at different distances from some pond weed in a test tube. They counted the number of bubbles produced by the plant. Here is the data they collected.

Distance from lamp in cm	5	10	15	20	25	30
Number of bubbles per minute	124	88	64	42	28	16

(a)   Plot a graph to show the results in the table.   **(2 marks)**

> **Maths skills**   Mark the points accurately on the grid (to within half a square) using the table of data. Then draw a line of best fit through these points. This line does not have to be straight.

(b)   Use your graph to find the number of bubbles you would expect in 1 minute if the lamp was placed 12 cm from the pond weed.

.... **(1 mark)**

(c)   Describe the relationship between light intensity and rate of photosynthesis.

....

.... **(2 marks)**

(d) (i)   State **one** safety step you should take.

....

.... **(1 mark)**

    (ii)   Explain **one** step you should take to ensure your results are reliable.

> Explain means you have to say what the step is and why you need to take that step.

....

....

.... **(2 marks)**

> **Guided**

(e)   Describe how you could use a light meter to improve the experiment.

You could use the light meter to measure the .... at each

distance and then plot a graph of ....

.... **(3 marks)**

Specialised plant cells

1 The diagram shows part of a plant tissue specialised for transport.

(a) State the name of this type of tissue.

.... **(1 mark)**

A

mitochondrion

B

vacuole

companion cell

sieve cell

> **Guided**

(b) Explain how the features labelled A and B are adapted to the function of this tissue.

A ....

....

B There is a small amount of cytoplasm so ....

.... **(4 marks)**

(c) Explain why companion cells have many mitochondria.

> Mitochondria supply energy. You need to give this information AND explain why companion cells need lots of energy.

....

....

.... **(2 marks)**

2 (a) State the name of the vessels used to transport water in plants.

.... **(1 mark)**

> **Guided**

(b) Describe **three** ways in which these vessels are adapted for their function.

1 The walls are strengthened with lignin rings to ....

....

2 ....

....

3 ....

.... **(3 marks)**

Transpiration

1   A student set up the following experiment to investigate transpiration.

20 18 16 14 12 10 8 6 4 2 0
cm

air bubble

(a)   State what is meant by the term **transpiration**.

....

.... **(2 marks)**

(b)   State which part of the plant regulates the rate of transpiration.

.... **(1 mark)**

(c)   Explain what happens to the air bubble if:

> Remember that in an **explain** question you need to say what happens and why.

(i)   a fan is started in front of the plant

....

.... **(2 marks)**

(ii)   the undersides of the leaves of the plant are covered with grease.

....

.... **(2 marks)**

2   (a)   Explain how the guard cells open and close.

....

....

.... **(3 marks)**

Guided

(b)   The stomata are open during the day but closed at night. Explain why, in very hot weather, plants wilt during the day but recover during the night.

The stomata are open during the day, so water is lost by .... faster

than it can be absorbed by the .... Water is lost from the vacuoles and

the plant wilts. At night, the stomata ....

.... **(3 marks)**

Translocation

1 (a) State what is meant by **translocation**.

....

.... **(1 mark)**

(b) What is the name of the plant tissue responsible for translocation?

☐ **A** phloem

☐ **B** xylem

☐ **C** meristem

☐ **D** mesophyll **(1 mark)**

2 (a) Describe how radioactive carbon dioxide can be used to show how sucrose is transported from a leaf to a storage organ such as a potato.

> Guided

Radioactive carbon dioxide is supplied to the ....

....

....

.... **(3 marks)**

(b) Explain what the effect would be if an inhibitor of active transport were applied to the leaf in this experiment.

> This question requires you to apply knowledge from other parts of the course. Make sure you describe what would happen and give an explanation.

....

....

.... **(2 marks)**

3 The table lists some of the structures and mechanisms involved in movement of water and sucrose in the plant. Put an X in each row of the table to show whether the structure or mechanism is involved in the transport of water or sucrose.

> You might need to revise transpiration on page 54 of the Revision Guide before answering this question.

Structure or mechanism	Transport of water	Transport of sucrose
Xylem		
Phloem		
Pulled by evaporation from the leaf		
Requires energy		
Transported up and down the plant		

(5 marks)

Water uptake in plants

1   Explain the effect on transpiration of the following:

   (a)   an increase in light intensity

   ....

   ....   **(2 marks)**

   (b)   an increase in temperature

   ....

   ....   **(2 marks)**

2   Some students investigated the rate at which water evaporated from leaves using this apparatus.

   The students measured how far the air bubble travelled up the capillary tube in 5 minutes with the fan on, and with the fan off. They found that the bubble moved 90 mm with the fan off and 130 mm with the fan on.

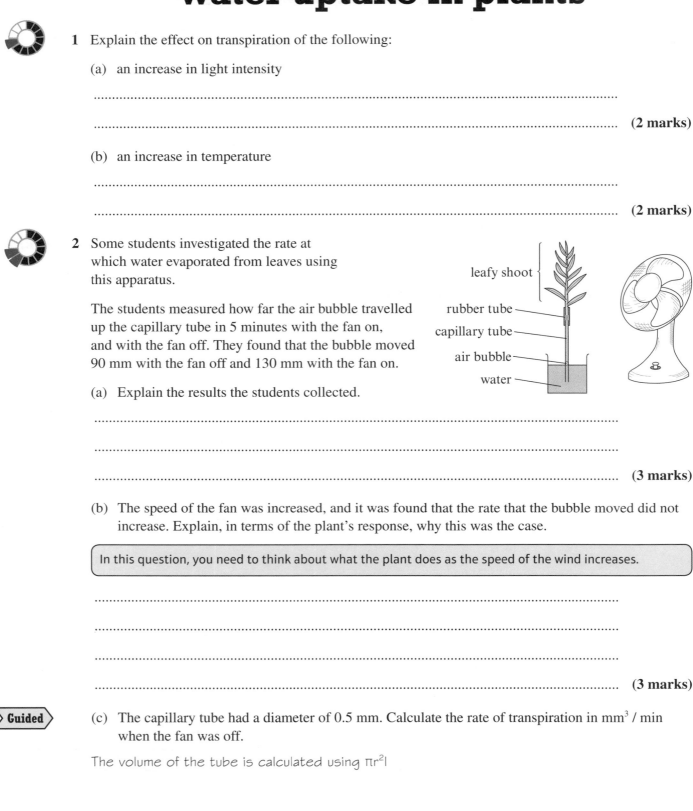

leafy shoot

rubber tube

capillary tube

air bubble

water

   (a)   Explain the results the students collected.

   ....

   ....

   ....   **(3 marks)**

   (b)   The speed of the fan was increased, and it was found that the rate that the bubble moved did not increase. Explain, in terms of the plant's response, why this was the case.

   > In this question, you need to think about what the plant does as the speed of the wind increases.

   ....

   ....

   ....

   ....   **(3 marks)**

   ⟩ **Guided** ⟩   (c)   The capillary tube had a diameter of 0.5 mm. Calculate the rate of transpiration in mm³ / min when the fan was off.

   The volume of the tube is calculated using πr²l

   rounded to 1 decimal place = ...   **(2 marks)**

Extended response – Plant structures and functions

Plant stomata are closed at night and open during the day. Explain how this process occurs, and how it provides leaves with materials needed for growth.

> You will be more successful in extended response questions if you plan your answer before you start writing.
>
> You need to explain:
>
> - how stomata open and close
> - the role of stomata in gas exchange
> - the importance of this in photosynthesis
> - the role of transpiration in providing mineral ions needed for growth.

....

....

....

....

....

....

....

....

....

....

....

....

.... **(6 marks)**

Hormones

1 (a) Describe how hormones behave like 'chemical messengers'.

Hormones are produced by .... and are released

into the .... They travel round the body until they reach

.... which responds by releasing

.... **(4 marks)**

(b) Describe **two** ways in which hormones and nerves communicate differently.

> Make sure you describe two ways and that they are differences, not similarities.

....

.... **(2 marks)**

2 The diagram shows the location of some endocrine glands in the body. Write in the name of each gland on the corresponding label line.

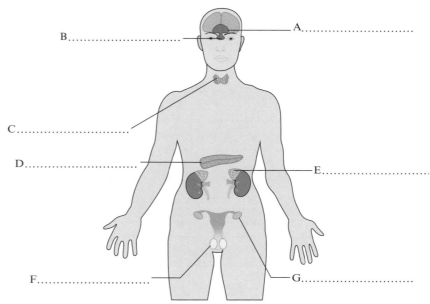

B...........................   A...........................

C...........................

D...........................   E...........................

F...........................   G...........................

(7 marks)

3 Complete the table showing where some hormones are produced and where they have their action.

Hormone	Produced in	Site of action
TRH		pituitary gland
TSH	pituitary gland	
ADH	pituitary gland	
FSH and LH		ovaries
insulin and glucagon		liver, muscle and adipose (fatty) tissue
adrenalin		various organs, e.g. heart, liver, skin
progesterone		uterus
testosterone		male reproductive organs

(4 marks)

Adrenalin and thyroxine

1 (a) A man is walking through a forest at dusk and hears a wolf howl. Explain **two** ways in which adrenalin prepares his body for action in this situation.

....

....

....

....

....

.... **(4 marks)**

(b) Thyroxine controls the resting metabolic rate. Explain how control of thyroxine concentration in the blood is an example of negative feedback.

> Note that in this question you are asked to talk about production of thyroxine, not how it works in controlling metabolic rate. Limit your explanation to the principles of negative feedback. The details will be needed in part (c).

....

....

....

.... **(2 marks)**

(c) Explain how the hypothalamus and pituitary work together to control the amount of thyroxine produced by the thyroid gland.

....

....

....

....

....

.... **(4 marks)**

2 The concentration of thyroxine in the blood is relatively constant but the concentration of adrenalin can change a lot. Explain the differences in the pattern of concentration of the two hormones in the blood.

Guided

Thyroxine controls the resting metabolic rate so ....

....

Adrenalin is produced in response to ....

....

....

.... **(4 marks)**

Had a go ☐   Nearly there ☐   Nailed it! ☐

The menstrual cycle

1 State **two** of the hormones that control the menstrual cycle.

.... Oestrogone ...and..progesturone .... **(2 marks)**

2 The diagram below shows the timing of some features in a menstrual cycle.

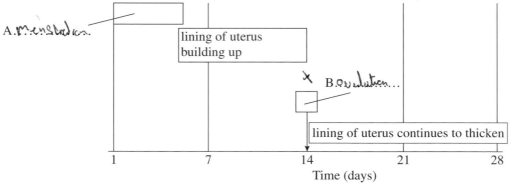

A. menstuden

lining of uterus building up

X B. Ovulation...

lining of uterus continues to thicken

1   7   14   21   28
Time (days)

(a) Fill in the two missing labels, A and B, on the diagram.   **(2 marks)**

(b) Mark with an X on the diagram the point at which fertilisation is most likely to occur.   **(1 mark)**

(c) Describe what happens during days 1–5 of the cycle.

...Menstrabuers - when the .....uterus lining..breaks...down........

....

And is lost in a bleed .... **(2 marks)**

3 (a) Explain how hormonal contraception prevents pregnancy.

Guided

Pills, implants or injections release hormones that prevent ..ovulation.......

and thicken .mucus in the cervix...., preventing ....Sperm....... from passing.   **(3 marks)**

(b) The table shows the success rates of different methods of contraception.

Method of contraception	Success rate (% of pregnancies prevented)
hormonal pill or implant	> 99%
male condom	98%
diaphragm or cap	92 – 96%

(i) Explain why the actual success rate can sometimes be lower than the figures shown.

..Not everyone would tell if..they..had..used a condom..but....
Still got pregnant .... **(2 marks)**

(ii) Evaluate the different methods of contraception for effectiveness in preventing pregnancy as well as protection against STIs.

> Be sure to answer both parts of this question. You might need to review the section on STIs on page 39 of the Revision Guide.

....A condom has a Success rate of 98% of pregnancies prevented....
..It also has a high percentage thing in preventing STIs....
..Another method of Contraption is the male Con pill which...
.has >99% Success at g prevent pregnancies whate it is very ...... **(3 marks)**
Weak at stopping STIs as there could still be a risk of b daily fluts

Control of the menstrual cycle

1 For each of the following, state where it is released and the target organ.

> This can be confusing, because some hormones are made in the pituitary and act on the ovary while others are made in the ovary and act on the pituitary. Make sure you get these the right way round.

(a) LH

....

.... **(2 marks)**

(b) progesterone

....

.... **(2 marks)**

(c) oestrogen

....

.... **(2 marks)**

Guided

2 The diagram shows how the levels of four hormones change during the menstrual cycle.

(a) (i) Explain why the level of FSH rises during the first 7 days of the cycle.

Levels of progesterone are low which

....

....

.... **(2 marks)**

(ii) Explain what causes ovulation to occur around day 14 of the cycle.

....

.... **(2 marks)**

(iii) Explain whether the female is pregnant.

....

.... **(2 marks)**

(b) Hormonal contraception uses a progesterone-like hormone. Explain how this can prevent fertilisation.

....

....

.... **(3 marks)**

(c) At the same time that the level of LH increases there is an increase in the woman's body temperature. Explain how a woman could use this to increase her chance of conceiving a child.

....

.... **(2 marks)**

Assisted Reproductive Therapy

> Guided

1 (a) Clomifene therapy can be useful for women who have difficulty conceiving a child. Explain how clomifene increases the chance of pregnancy in women who rarely release an egg during their menstrual cycles.

Clomifene helps increase the concentration of ....

so stimulates ....

.... **(2 marks)**

(b) Explain why clomifene therapy on its own cannot help a woman with blocked oviducts to become pregnant.

> Think about what clomifene does and why it might not help a woman with blocked oviducts.

....

....

.... **(2 marks)**

2 IVF can be used to help couples who are unable to conceive a child because of problems such as blocked oviducts in the woman or if the man produces few healthy sperm cells.

(a) Explain why the woman is given injections of the hormone FSH at the start of IVF treatment.

....

....

.... **(2 marks)**

(b) Describe the steps that take place after egg cells from the woman and sperm cells from the man are obtained.

....

....

.... **(2 marks)**

(c) Explain why IVF can be useful for couples who risk passing on genetic disorders even if they are able to conceive normally.

....

....

.... **(2 marks)**

3 In 2010, 45 250 women underwent IVF treatment in the UK. Of these women, 12 400 were successful in having a child. The cost of a cycle of IVF treatment is £2500.

Use this information, and your own knowledge, to describe the benefits and drawbacks of IVF treatment.

....

....

....

....

.... **(4 marks)**

Blood glucose regulation

1   The table shows the events that happen after a person eats a meal. Complete the table to show the order in which the events take place.

Event	Order
Pancreas increases secretion of insulin	
Blood glucose concentration falls	
Blood glucose concentration rises	1
Insulin causes muscle and liver cells to remove glucose from blood and store it as glycogen	
Pancreas detects rise in blood glucose concentration	

(3 marks)

2   (a)   The diagram shows how blood glucose concentration is regulated. Use the information below to fill in the corresponding boxes on the diagram.

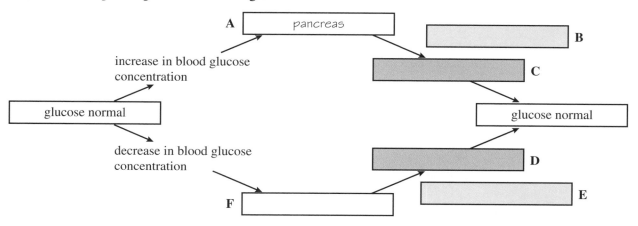

A and F – this gland releases hormones that regulate blood glucose concentration

B – the name of a hormone involved in regulating blood glucose concentration

C and D – the target organ for the hormones involved in blood glucose regulation

E – the name of a hormone involved in regulating blood glucose concentration   **(4 marks)**

(b)   Explain what happens in the liver when:

(i)   the blood glucose concentration rises above normal

....

....

....

....   **(2 marks)**

(ii)   the blood glucose concentration drops below normal.

....

....

....

....   **(2 marks)**

Diabetes

1 (a) The graph shows the percentage of people in one area of the USA in the year 2000 who have Type 2 diabetes, divided into groups according to body mass index.

Describe the link between BMI and the percentage of Type 2 diabetes.

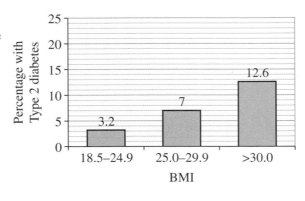

....

.... **(2 marks)**

(b) Two 45-year-old males from the area of the USA studied in part (a) wanted to estimate their chances of developing Type 2 diabetes.

(i) George was 180 cm tall and had a mass of 88 kg. Calculate his BMI and use this to evaluate his risk of developing Type 2 diabetes.

.... **(3 marks)**

(ii) Donald had a waist measurement of 104 cm and a hip measurement of 102 cm. The World Health Organization classes a waist : hip ratio of >0.9 as obese. State and explain whether Donald has an increased risk of developing Type 2 diabetes.

....

....

.... **(3 marks)**

2 (a) Explain how helping people to control their diets might help to reduce the percentage of people in the population who have diabetes.

Controlling diets will help to ....

Fewer obese people means .... **(2 marks)**

(b) (i) Explain why people with Type 1 diabetes are treated with insulin but most people with Type 2 diabetes are not.

....

.... **(2 marks)**

(ii) Explain why a person with Type 1 diabetes will sometimes wait to see how large a meal is before deciding how much insulin to inject.

....

.... **(2 marks)**

Extended response –
Control and coordination

 Compare how Type 1 and Type 2 diabetes are caused and how they are treated.

> You will be more successful in extended response questions if you plan your answer before you start writing.
>
> Make sure that you cover the causes of each type of diabetes and link this to type of treatment.

....

....

....

....

....

....

....

....

....

....

....

....

.... **(6 marks)**

Exchanging materials

1 Substances are transported into and out of the body. Describe where and why the following substances are transported in this way.

(a) Water .... **(2 marks)**

(b) Urea .... **(2 marks)**

2 Dissolved food molecules are transported into the body from the digestive system. Describe where and why these substances are transported in this way.

....

....

.... **(2 marks)**

3 Humans and other mammals need to exchange gases with their environment. Describe where and why this exchange happens.

....

....

.... **(3 marks)**

4 Absorption of digested food molecules takes place in the small intestine. The small intestine has a surface adapted to assist this process.

Guided

(a) Describe how the small intestine is adapted to help to absorb food molecules.

The surface of the small intestine is covered with .... These help

by increasing .... **(2 marks)**

(b) Explain why the structures described in part (a) have thin walls.

....

.... **(2 marks)**

5 The diagram shows a flatworm and an earthworm.

Guided

The two worms are similar in size. Explain why the flatworm does not have an exchange system or a transport system whereas the earthworm has a transport system (heart and blood vessels).

The flatworm is very flat and thin which means it has a large ....

....

....

....

....

.... **(4 marks)**

Alveoli

1 (a) Describe how gas exchange takes place in the lungs.

Guided

Oxygen diffuses from ....*air*........ into ....*blood*........

Carbon dioxide diffuses from ....*Blood*........ into ....*Alveoli Air*........ **(2 marks)**

Guided

(b) State and explain **two** ways in which the structure of the alveoli is adapted for efficient gas exchange.

Millions of alveoli create a large ....*Surface area*........

for the ....*diffusion*........ of gases. Each alveolus is closely associated with

a ....*Capillary*........ Their walls are one ....*cell thick*........

This minimises the ....*diffusion*........ distance. **(4 marks)**

2 Explain the importance of continual breathing and blood flow for gas exchange.

........*It maintains the high concentration gradients, to maximise rate of diffusion*........ **(2 marks)**

3 Emphysema is a type of lung disease where elastic tissue in the alveoli breaks down. The figure shows the appearance of an alveolus damaged by lung disease compared with a healthy alveolus.

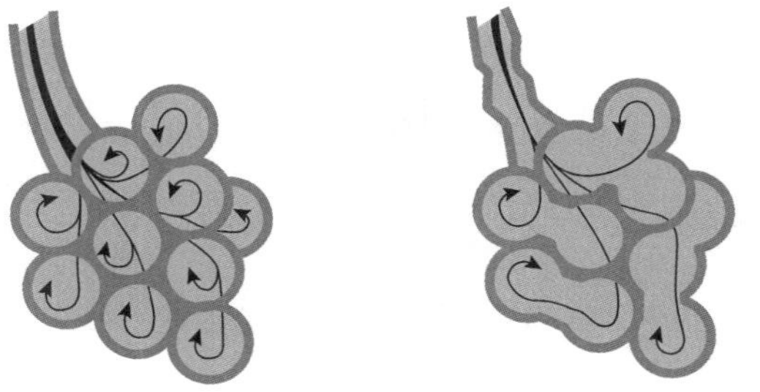

Healthy alveolus   Alveolus damaged by lung disease

Explain how emphysema affects the person.

> Think about what effects the changes in emphysema would have on gas exchange and how this would then affect the person.

........*This reduces the surface area of the alveoli over which diffusion can take place and so it decreases rate of gas exchange this causes the person to have a troubled time breathing.*........ **(3 marks)**

Blood

1 (a) Explain why red blood cells contain large amounts of haemoglobin.

haemoglobin carries oxygen which ~~means~~ fufuls
the purpose of rbc

(2 marks)

(b) Explain **two** other ways in which the structure of red blood cells is related to their function.

> Make sure that you give two features of red blood cells and, for each one, relate the structure to the function.

the biconcave structure means that they have
more surface area for haemoglobin meaning that
its easier for oxygen to diffuse in and out of the cell.

(4 marks)

2 Describe **one** way in which blood plasma transports substances.

Dissolved substances such as Carbon and glucose
are transported through the blood vessels

(2 marks)

3 Explain how platelets are adapted to protect the body from infection.

Platelets clot up to form a scab which
acts as a barrier for the wound which stops
pathogens getting into the blood

(3 marks)

4 White blood cells usually make up about 1% of the blood and include lymphocytes and phagocytes.

(a) Explain why the number of lymphocytes increases during infection.

lymphocytes produce antibodies that attach to pathogens
and destroy them.

(3 marks)

(b) Describe how phagocytes help protect the body.

white blood cells e.g phagocytes surround pathogens
and destroy them

(2 marks)

Blood vessels

1 (a) Describe the structure of an artery.

Guided

An artery has .... walls. These walls are composed of two types of fibres:

.... tissue and .... fibres.   **(3 marks)**

(b) Explain how the structure of the artery wall makes blood flow more smoothly in arteries.

....

....

....   **(2 marks)**

2 Blood needs to penetrate every organ in the body. This is made possible by capillaries.

(a) Describe how the capillaries are adapted for this function.

> Make sure you describe here and save the explanation for part (b).

....

....

....   **(2 marks)**

(b) Explain how the features you have described are important for the function of capillaries.

....

....

....   **(2 marks)**

3 (a) Veins carry blood away from the organs of the body to the heart.

(i) Explain why there is a difference in the thickness of the walls of arteries and veins.

....

....

....   **(2 marks)**

(ii) Explain how muscles and valves work together to help to return blood to the heart.

....

....

....   **(2 marks)**

(b) A nurse taking blood from a patient will insert a needle into a vein.

Explain why blood is taken from veins, not from arteries.

> There are two possible answers here – you need to consider either the structure of the different blood vessels, or else the way in which each transports the blood they contain.

....

....

....   **(2 marks)**

The heart

Guided

1 The heart is connected to four major blood vessels. Describe where each vessel carries blood. The first one has been done for you.

aorta   carries blood from heart to body

pulmonary artery   carries blood from ...heart........... to ...lungs...........

pulmonary vein   carries blood from ...lungs........... to ...heart...........

vena cava   carries blood from ...body........... to ...heart........... **(4 marks)**

2 (a) Explain why the heart consists mostly of muscle.

......The heart must have muscle as it has to.....

......pump blood around the whole body..................... **(2 marks)**

(b) Describe the route taken by blood through the heart from the vena cava to the aorta.

....

....

.... **(3 marks)**

3 The diagram shows a section through the human heart.

> Remember that the heart is drawn and labelled as if you are looking at the heart in someone's body. So the right side of the heart is actually on the left side of the page!

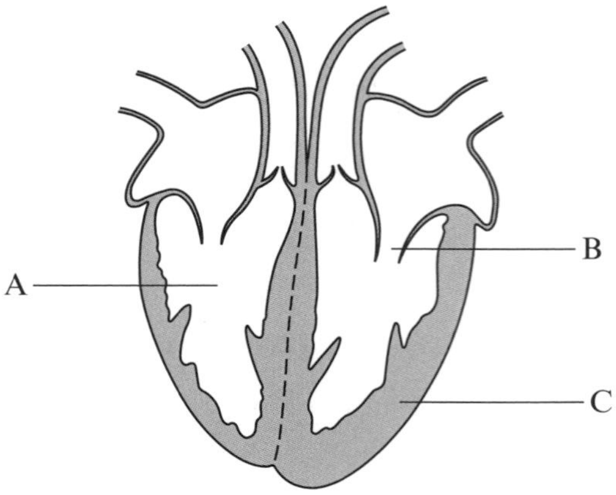

A —

B

C

(a) State the name of the part of the heart labelled A and describe its function.

....

.... **(2 marks)**

(b) Explain the function of the part labelled B.

....

.... **(2 marks)**

(c) Explain why the muscle at C needs to be thicker than on the other side of the heart.

....

....

.... **(3 marks)**

Aerobic respiration

1 Read the following passage and answer the questions that follow.

> Aerobic respiration happens in muscle cells in the body. The muscle cells are surrounded by blood vessels. The substances needed for respiration are transferred to the muscle cells by diffusion, and the waste products are removed.

(a) Name the substances needed for respiration in muscle cells.

............Glucose............ and ....Oxygen.......... **(2 marks)**

Guided

(b) State the meaning of the term **diffusion**.

Diffusion is the movement of substances from ....High.... to ....low...../concentration.

(1 mark)

2 (a) State the location in the cell where most of the reactions of aerobic respiration occur.

..........Mitochondria.......... **(1 mark)**

(b) Explain how cellular respiration helps maintain the body temperature.

....It releases energy which could contain thermal So It releases temperature

(2 marks)

(c) State **one** way that animals use energy from respiration, other than to maintain their body temperature.

.......to enable muscle contraction.......

(1 mark)

3 The blood supplies cells with the substances needed for aerobic respiration, as well as removing waste products.

(a) Write a word equation for aerobic respiration.

.....Glucose + oxygen → carbon dioxide + water (1 mark)

(b) State the name of the smallest blood vessels that carry blood to the respiring cells.

.....Capillery........ **(1 mark)**

4 (a) Explain why all organisms respire continuously.

....To mantain a Steady temperature.......... ✗

(2 marks)

(b) Plants can use energy from sunlight in photosynthesis. Explain why plants also need to respire continuously.

> Photosynthesis uses light energy in the production of glucose; it does not release energy that can be used in other processes. Think about why plants need energy from respiration.

....To produce energy and be alive.......... ✗

(2 marks)

71

Anaerobic respiration

1 Humans can respire in two ways: using oxygen (aerobic) and without using oxygen (anaerobic).

> Make sure that you understand what is produced in both aerobic and anaerobic respiration.

(a) Compare the amounts of energy transferred by aerobic and anaerobic respiration.

anerobic respiration releases less energy than aerobic. **(2 marks)**

(b) Describe the circumstances under which anaerobic respiration occurs.

A when energy is needed for muscle contraction if there is not enough oxygen **(2 marks)**

2 In track cycling, a 'sprint' event begins with several slow laps in which the riders try to get a tactical advantage. These slow laps are followed by a very fast sprint to the finishing line.

(a) Describe and explain how the cyclists' heart rates change during the course of the race.

....

....

.... **(3 marks)**

(b) After the race the cyclists will cycle on a stationary bicycle for 5–10 minutes. Explain why they do this.

This is so they can get all the oxygen back

.... **(2 marks)**

3 The graph shows how oxygen consumption changes before, during and after exercise. The intensity of the exercise kept increasing during the period marked 'Exercise'.

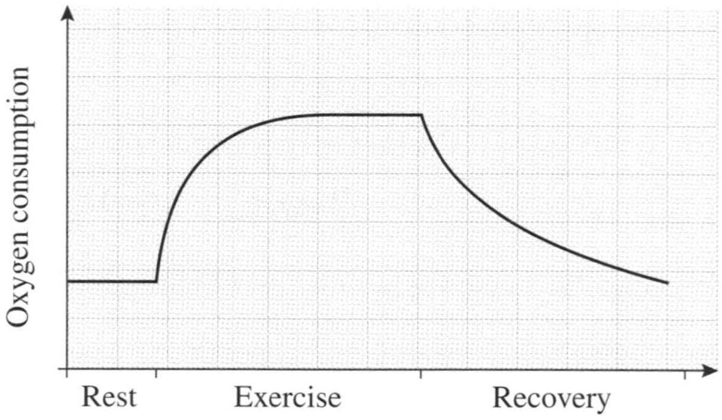

(a) Explain the shape of the graph during the period marked 'Exercise'.

....

.... **(3 marks)**

Guided

(b) Explain the shape of the graph during the period marked 'Recovery'.

During exercise there is an increase in the concentration of *blood lactic*

acid this is because aerobic respiration can't supply enough energy so anaerobic respiration means products build up **(2 marks)**

 **Practical skills**

Rate of respiration

1   The diagram shows a respirometer used to investigate the rate of respiration in germinating peas.

syringe containing air

3-way tap

ruler

capillary tubing   blob of liquid

respiring peas
wire gauze
water bath
potassium hydroxide

> This is one of the core practicals so you should be able to answer questions on the apparatus.

State the role of the following and give a reason for your answer.

(a) the water bath

....

....

.... **(2 marks)**

> **Guided**

(b) the potassium hydroxide

absorbs carbon dioxide produced by the .... so that

....

.... **(2 marks)**

(c) the tap and syringe containing air

....

....

.... **(2 marks)**

2   (a) Explain how the apparatus allows you to measure the rate of respiration in the seeds.

> The movement of the liquid blob gives you information about the uptake of oxygen. You need to state this, explain how you measure the movement and then how you calculate the rate of respiration.

....

....

.... **(3 marks)**

(b) Describe how you would use the apparatus to investigate the effect of temperature on the rate of respiration in peas.

....

....

.... **(3 marks)**

Changes in heart rate

1   Cardiac output can be calculated using the equation:
cardiac output = stroke volume × heart rate.

(a)   What is meant by the term **stroke volume**?

.... **(1 mark)**

(b)   A man has a heart rate of 60 beats/minute and an average stroke volume of 75 cm³.

(i)   Calculate his cardiac output. Show your working out and give the correct units.

cardiac output .... **(3 marks)**

(ii)   Explain the change in cardiac output when the man starts to exercise.

....

....

.... **(3 marks)**

2   The graph shows the pulse rate of an athlete at rest, and after 5 minutes of different types of exercise.

Remember to show all your steps in the calculation.

(a)   Calculate the percentage increase in pulse rate between jogging and running.

100 beats/min – 80 beats/min = ...beats/min

(.........../80) × 100 =

percentage increase .... **(2 marks)**

(b)   State why the pulse rate is highest when the athlete is rowing.

....

.... **(1 mark)**

(c)   The pulse is a measure of heart rate. At rest, the cardiac output of the athlete is 4000 cm³/min.
Calculate the stroke volume, in cm³, of the athlete at rest.

stroke volume .... cm³ **(2 marks)**

Extended response – Exchange

The diagram shows the main features of the human heart and circulatory system.

Describe the journey taken by blood around the body and through the heart, starting from when it enters the right side of the heart. In your answer, include names of major blood vessels and chambers in the heart.

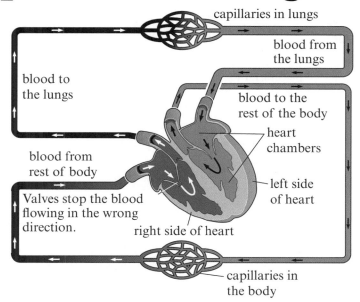

capillaries in lungs

blood from the lungs

blood to the lungs

blood to the rest of the body

heart chambers

left side of heart

blood from rest of body

Valves stop the blood flowing in the wrong direction.

right side of heart

capillaries in the body

You will be more successful in extended response questions if you plan your answer before you start writing.

You do not need to explain how the different components of the heart and circulatory system work. It may help your plan if you follow the blood around the diagram with a finger, writing the name of each blood vessel or chamber in order as you go. You do not need to identify any blood vessels in the 'rest of the body' other than the aorta.

....

....

....

....

....

....

....

....

....

....

....

....

.... **(6 marks)**

Ecosystems and abiotic factors

1 Draw lines to connect each term with its definition.

Term	Definition
Community	A single living individual
Organism	All the living organisms and the non-living components in an area
Population	All the populations in an area
Ecosystem	All the organisms of the same species in an area

(4 marks)

2 A student surveyed the distribution of a species of lichen growing on the trunk of a tree. She used a small quadrat to measure the percentage cover by these lichens on the south and north facing sides of the tree.

	Light intensity (lux)			
	Reading 1	Reading 2	Reading 3	Mean
South side	275.5	368.1	326.8	
North side	195.7	282.1	205.1	

Percentage cover (%)											
South side						North side					
1	2	3	4	5	Mean	1	2	3	4	5	Mean
48	20	28	92	8	39	4	4	4	4	6	4

Guided

(a) Complete the upper table to show the mean light intensity for each side. **(2 marks)**

(275.5 + 368.1 + 326.8)/3 = ...

(b) The student concluded that the lichen was better adapted to conditions on the south side. Justify her conclusions.

> In this part of the question you are only asked to say whether she was right and, if so, why. Don't try to explain her results.

....

....

....

.... **(2 marks)**

(c) Light intensity is an abiotic factor. Explain one other abiotic factor that might be responsible for the different distribution of lichen.

....

....

....

.... **(2 marks)**

Biotic factors

1   Meerkats are animals that live in packs and are found in the desert areas of southern Africa. The pack of meerkats is led by a dominant pair of meerkats, known as the alpha male and female.

> In this question, you will be asked to think about aspects of the behaviour of the meerkats. Remember to link your answer to the ideas that these animals will compete with each other for resources.

(a)   State what the term **biotic factors** means.

.... **(1 mark)**

(b)   Only the alpha male and alpha female breed. Suggest an explanation for why younger male meerkats will often try to fight the alpha male.

....

.... **(2 marks)**

(c)   When meerkat packs become very large, they often split into smaller packs. The new pack will often move some distance from the original pack. Explain the reasons why a large pack may need to split up.

> Make sure that you know what the command words mean. **Explain** means give a reason why. **Suggest** means you need to apply your knowledge to a new situation. **Describe** means say what is happening.

....

.... **(2 marks)**

2   The drawing shows a male peacock.

State and explain **one** adaptation, seen in the diagram, that helps the peacock attract a mate.

....

....

....

.... **(3 marks)**

3   The diagram shows a cross-section through a tropical rainforest.

Guided

(a)   Some trees are called emergent. They break through the rest of the rainforest canopy. Explain the advantage to these trees of emerging from the canopy.

The trees emerge through the canopy

to get .... for more ....

(2 marks)

emergent layer

canopy layer

(b)   The soil in a rainforest is often poor as the minerals are washed away (leached). Suggest an explanation of how trees in the rainforest may adapt to respond to a leached soil.

....

.... **(2 marks)**

Parasitism and mutualism

1 Compare and contrast parasitism and mutualism.

....

....

.... **(3 marks)**

> You could get a question like this as an extended response question in an exam, where you would probably be asked to give examples to illustrate your answer. Here it is enough to say how the relationships are similar and how they are different.

2 Fleas are small insects that feed on the blood of animals.

(a) Describe what each organism gets out of this relationship.

Fleas: ....

....

Animals: ....

.... **(2 marks)**

(b) Explain what type of feeding relationship exists between fleas and animals.

....

....

.... **(2 marks)**

3 Cleaner fish are small fish that feed on parasites on the skin of sharks. Describe how the cleaner fish and the sharks benefit from a mutualistic relationship.

> **Guided**

Cleaner fish get food by..

....

This helps the shark because .... **(2 marks)**

> An exam question may ask you about the benefits to one organism or to both. Make sure you read the question carefully!

4 The scabies mite is a tiny insect that burrows into human skin and lays its eggs. Infection by the scabies mite causes severe itching and a lumpy, red rash that can appear anywhere on the body. Explain why the scabies mite is a parasite and not a mutualist.

> This type of question is expecting you to apply your understanding of science to a situation that you may not be familiar with. You will have been taught about organisms that behave in a similar way to the scabies mite – use what you know about these organisms but apply it to the scabies mite.

....

....

.... **(2 marks)**

Fieldwork techniques

1   A gardener goes into his garden every night at 7 pm and counts the number of slugs in the same 1 m² area of his flower bed. He records his results in a table.

Day	Monday	Tuesday	Wednesday	Thursday	Friday	Saturday	Sunday
Number of slugs	11	12	7	12	8	8	12

(a)   Describe how the gardener could make sure the 1 m² area of the flower bed was chosen at random on the first day.

....

.... **(2 marks)**

Guided

(b)   Why does the gardener use the same area each time?

Using the same area means that his experiment is .... **(1 mark)**

(c)   Describe **one** way in which the gardener could improve the repeatability of the data that he collected.

....

.... **(2 marks)**

2   A class is investigating the number of clover plants on a football pitch. The pitch measures 100 m by 65 m. The class wants to find the total number of clover plants in the field. The teacher gives the class a 1 m × 1 m quadrat.

> **Maths skills**   mean number of plants = $\dfrac{\text{total number of plants in all quadrats}}{\text{number of quadrats}}$

(a)   Explain how the class can use the quadrat to estimate the mean number of clover plants in a 1 m² area.

....

.... **(2 marks)**

(b)   The class finds that the mean number of clover plants in an area of 1 m × 1 m is 7. Estimate the number of clover plants on the whole football pitch.

(3 marks)

3   Describe how you would use a belt transect to investigate the distribution of broad-leaved plants growing alongside a path that started at a road, crossed a small field and entered a wood.

> **Practical skills**   Make sure you describe use of quadrats, the measurements you would take and what you would record.

....

....

....

.... **(3 marks)**

Organisms and their environment

Practical skills

1   Limpets are animals that have a shell and live on rocks that are underwater some or all of the time. They can be found in the sea, or in rock pools on the beach. A scientist is investigating the distribution of limpets on the beach.

(a)   Explain how the scientist could use a transect to investigate the distribution of limpets.

....

....

.... **(3 marks)**

The scientist sets up three different transects and measures the numbers of limpets on each one. His data is shown in the table.

Distance from sea in metres	Number of limpets			Mean number of limpets
	Transect 1	Transect 2	Transect 3	
0.5	20	23	20	21
1.0	18	16	17	17
1.5	13	13	13	13
2.0	10	8	9	
2.5	5	6	4	5

(b)   Calculate the mean number of limpets at 2.0 m from the sea in this investigation.

.... **(2 marks)**

(c)   What conclusion can be made from his investigation?

> Your conclusion should describe how the distribution of limpets changes along the transect.

....

.... **(3 marks)**

2   A scientist investigated the distribution of bluebells in a large wood. She started on the edge of the wood, and measured a line going deeper into the wood. Every 2 m into the woodland, she placed a quadrat and counted the number of bluebells in the quadrat. She also measured the light intensity at each quadrat.

Guided

(a)   Describe **one** way the scientist could alter her method to collect more accurate data.

Instead of placing a quadrat every 2 m, the scientist could ....

.... and use a .... quadrat than before.   **(2 marks)**

(b)   The scientist obtained the following data:

Distance from edge of wood in metres	0	2	4	6	8	10	12	14	16
Number of bluebells	0	7	15	22	25	21	16	10	8

Suggest an explanation for these results.

....

....

....

.... **(2 marks)**

Human effects on ecosystems

1   Fish can be farmed or caught from the wild. State **one** advantage of fish farming, and **one** disadvantage.

> **Guided**

Advantage Reduces fishing of ....

Disadvantage ....

.... **(2 marks)**

2   A non-indigenous species is not naturally found in a particular place. For example, the cane toad is a non-indigenous species in Australia that was introduced to control insect pests. State **one** other advantage of introducing a non-indigenous species, and **one** disadvantage.

Advantage ....

....

Disadvantage ....

....

.... **(2 marks)**

3   The graph shows the mass of fertiliser used in the world from 1950 to 2003.

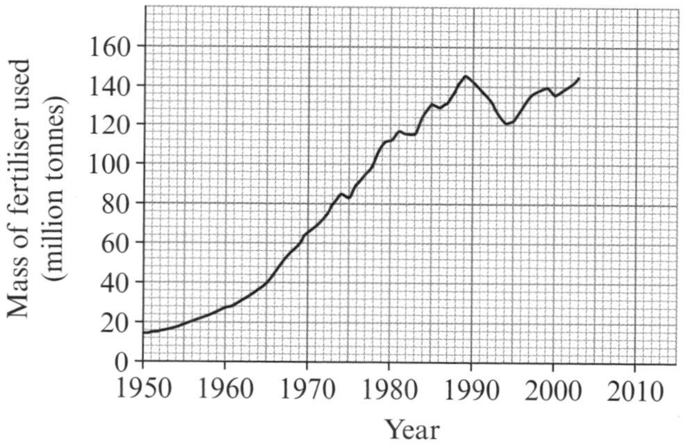

(a)   Calculate the percentage increase in fertiliser use from 1950 to 2003.

> Make sure that you read the graph carefully to get the correct figures for your calculation. You will get one mark for showing the correct calculation and one mark for the correct answer.

percentage increase.. **(2 marks)**

(b)   Suggest an explanation for the change in the mass of fertiliser used worldwide since 1950.

....

....

.... **(2 marks)**

(c)   Describe an environmental problem caused by over-use of fertilisers.

....

.... **(2 marks)**

Biodiversity

1 (a) State what is meant by **reforestation**.

....

.... **(1 mark)**

(b) Describe **two** advantages of reforestation.

....

....

.... **(2 marks)**

2 Explain the importance to humans of conservation.

....

....

.... **(2 marks)**

Guided

3 Yellowstone National Park in the USA is the natural home of many species. Deer eat young trees, stopping them from growing. The population of deer increased so much that Park authorities decided to reintroduce some wolves.

The wolves killed some of the deer for food. The deer moved away from river areas because they were more easily hunted by the wolves there. The wolves also killed coyotes, which are predators that eat rabbits.

The reintroduction of wolves led to major improvements in the biodiversity of the Park. This included increases in the populations of rabbits, bears, hawks and other birds. It also reduced the amount of soil washed into the rivers.

Describe the ways in which the reintroduction of the wolves may have caused the biodiversity to improve.

> This is an example of having to apply knowledge you have learned in this and other units to an unfamiliar situation.

The numbers of trees will increase because ....

This means there will be more food for ....

There will be more rabbits because ....

If there are more rabbits, there will be more food for . ...

More trees also mean ....

.... **(6 marks)**

The carbon cycle

1 Complete the diagram of the carbon cycle by writing the names of the processes in the boxes.

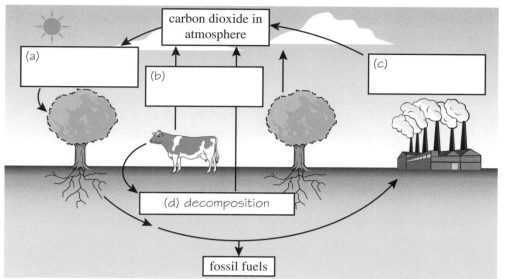

(4 marks)

2 Explain why microorganisms are important in recycling carbon in the environment.

....

....

.... **(2 marks)**

> In questions about the carbon cycle, you will be expected to make links between
> photosynthesis, respiration and combustion, and the amount of carbon dioxide in the air.

3 (a) The diagram shows a fish tank. Explain how
carbon is recycled between organisms in the
fish tank.

....

....

....

....

....

....

.... **(4 marks)**

(b) Explain why it is important that numbers of both plant and animal populations in the
fish tank are kept balanced.

....

....

.... **(3 marks)**

The water cycle

1 (a) Give **three** natural sources of water vapour in the atmosphere.

> The question says 'water vapour' so make sure your answer talks about the formation of water vapour and not about other aspects of the water cycle.

....

....

....

....

.... **(3 marks)**

(b) Describe what happens when water vapour in the atmosphere condenses.

....

....

....

....

.... **(3 marks)**

Guided

2 In parts of California there is a lack of rainfall. Water has been taken from rivers and used to water lawns and golf courses. Some of these areas are suffering from drought and there are now restrictions on the number of days a week golf courses can be watered. Explain why these restrictions have been introduced.

A lot of water evaporates from a golf course so this will lead to.................................

....

....

.... **(3 marks)**

3 Sea water contains too much salt to make it potable (safe to drink). Potable water can be produced from sea water by desalination:

• sea water is evaporated by heating

• water vapour is cooled and condensed.

Give **one** advantage of desalination to people in areas where there is a drought, and **one** disadvantage.

....

....

....

....

....

.... **(2 marks)**

The nitrogen cycle

The diagram shows how the element nitrogen moves between living organisms and the environment.

1   Bacteria are involved in different stages of the nitrogen cycle. Which is the correct combination of processes involving bacteria?

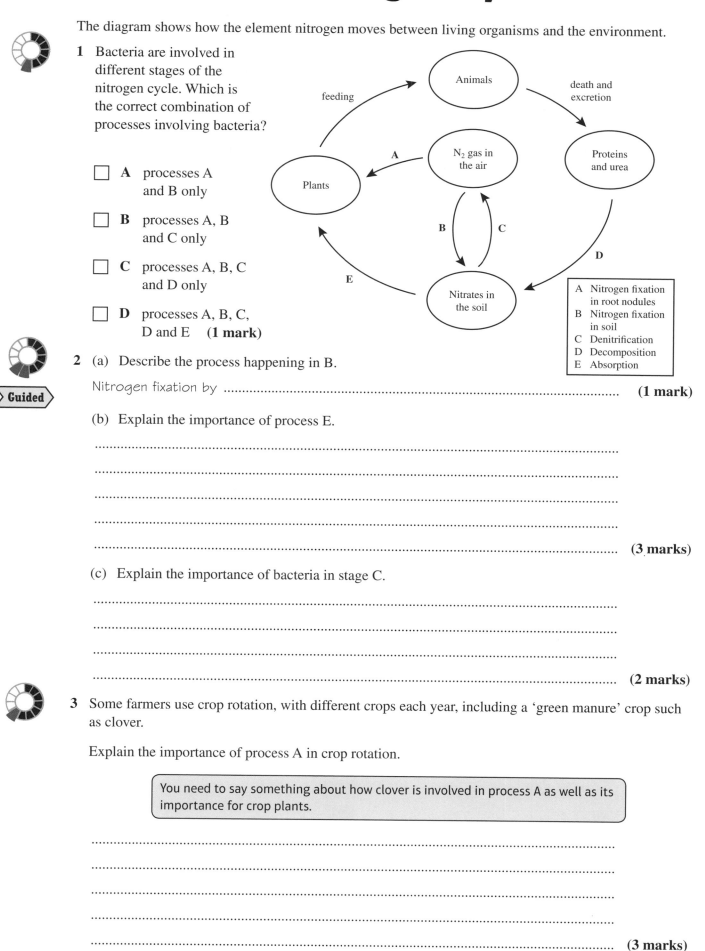

☐ **A**   processes A and B only

☐ **B**   processes A, B and C only

☐ **C**   processes A, B, C and D only

☐ **D**   processes A, B, C, D and E   **(1 mark)**

A   Nitrogen fixation in root nodules
B   Nitrogen fixation in soil
C   Denitrification
D   Decomposition
E   Absorption

2   (a)   Describe the process happening in B.

> **Guided**

Nitrogen fixation by .... **(1 mark)**

(b)   Explain the importance of process E.

....

....

....

....

.... **(3 marks)**

(c)   Explain the importance of bacteria in stage C.

....

....

....

.... **(2 marks)**

3   Some farmers use crop rotation, with different crops each year, including a 'green manure' crop such as clover.

Explain the importance of process A in crop rotation.

> You need to say something about how clover is involved in process A as well as its importance for crop plants.

....

....

....

....

.... **(3 marks)**

Extended response – Ecosystems and material cycles

Explain how fish farming and other human activity has an impact on biodiversity.

> You will be more successful in extended response questions if you plan your answer before you start writing.
>
> Try to include a number of different examples of how human activity has an impact on biodiversity. Remember that not all human activity is bad for biodiversity; try to think of some examples where human activity can increase biodiversity.

....

....

....

....

....

....

....

....

....

....

....

....

.... **(6 marks)**

Formulae

1   Which of the following is the formula for calcium carbonate?

☐ **A** CaCO

☑ **B** $CaCO_2$

☐ **C** $CaCO_3$

☐ **D** $CaCO_4$ **(1 mark)**

2   State what is meant by the term **element**.

An element is a substance made from ...*atoms*...

with the same number of...*protons*... **(2 marks)**

3   Chlorine is used to kill harmful microorganisms in drinking water. Its formula is Cl_2.

(a) Explain, using the information given, how you know that chlorine is **not** a compound.

...*chlorine isnt chemically joined*...........................
...*together but Cl_2 is two elements*................
...*in one*... **(2 marks)**

(b) Explain, using the information given, how you can tell that chlorine exists as molecules.

...*A molecule is two or more atoms*................
...*chemically bond together*.....................................
.... **(2 marks)**

4   Complete the table to show the formulae of some common substances.

Substance	water	carbon dioxide	methane	sulfuric acid	sodium
Formula					

(5 marks)

5   The formula for aluminium hydroxide is $Al(OH)_3$.

(a) State the number of elements in the formula $Al(OH)_3$.

...*Aluminium oxygen hydrogen*................................... **(1 mark)**

(b) State the total number of atoms in the formula $Al(OH)_3$.

...*7*... **(1 mark)**

6   The formula for a carbonate ion is CO_3^{2-}. Describe what this formula shows.

...*The ion has 2 negative charges and is*..............
...*formed from one carbon atom and 3 oxygen chemically*......... **(2 marks)**
bined together

Equations

1 Which of these statements describes a chemical reaction?

☐/ **A** Reactants form from products.

☑ **B** Products form from reactants.

☐ **C** An element changes into another element.

☐ **D** The total mass of substances goes down.

> Answer **C** cannot be correct because one element cannot change to another element in chemical reactions.

(1 mark)

2 Sodium hydroxide solution reacts with dilute hydrochloric acid to form sodium chloride and water.

(a) Write the word equation for this reaction.

Sodium Hydroxide$_{(l)}$ + hydrochloric acid$_{(l)}$ → Sodium chloride + water

(1 mark)

(b) Write the balanced equation for this reaction.

2NaOH + 2HCl → 2NaCl + 2H$_2$O

(1 mark)

3 A teacher adds a piece of sodium metal to some water. The reaction produces sodium hydroxide solution and bubbles of hydrogen.

> You should know the formulae of elements and simple compounds.

> Guided

(a) Complete the balanced equation below to show the correct state symbols.

$2Na(s) + 2H_2O(..l....) \rightarrow 2NaOH(..aq..) + H_2(..g..)$

(1 mark)

(b) Describe how you know that the equation above is balanced.

All are equal

(2 marks)

4 The following equations are **not** balanced. Write the balanced equations in the spaces below them.

> Do not add state symbols unless you are asked for them.

(a) $Cu + O_2 \rightarrow CuO$

2Cu + O$_2$ → 2CuO

(1 mark)

(b) $Al + Fe_2O_3 \rightarrow Al_2O_3 + Fe$

2Al + Fe$_2$O$_3$ → Al$_2$O$_3$ + 2Fe

(1 mark)

(c) $Mg + HNO_3 \rightarrow Mg(NO_3)_2 + H_2$

Mg + 2HNO$_3$ → Mg(NO$_3$)$_2$ + H$_2$

(1 mark)

(d) $Na_2CO_3 + HCl \rightarrow NaCl + H_2O + CO_2$

Na$_2$CO$_3$ + 2HCl → 2NaCl + H$_2$O + CO$_2$

(1 mark)

(e) $Fe + O_2 \rightarrow Fe_2O_3$

4Fe + 3O$_2$ → 2Fe$_2$O$_3$

(1 mark)

(f) $Cl_2 + NaBr \rightarrow NaCl + Br_2$

Cl$_2$ + 2NaBr → 2NaCl + Br$_2$

(1 mark)

Ionic equations

Guided

1 Explain what is meant by the term **ion**.

An ion is a *electrically charged particle*

formed when *an atom loses or gains electrons* **(2 marks)**

2 Silver nitrate solution is used to identify iodide ions in solution. A yellow precipitate of silver iodide, AgI, forms if iodide ions are present.

(a) Give the formula of the silver ion and the formula of the iodide ion in silver iodide.

silver ion ....

iodide ion .... **(2 marks)**

(b) Write the balanced ionic equation for the formation of silver iodide. Include state symbols in your answer.

.... **(2 marks)**

3 Dilute acids contain hydrogen ions. These react with carbonate ions to form water and carbon dioxide.

(a) Give the formula for a hydrogen ion and the formula for a carbonate ion.

hydrogen ion ....

carbonate ion.. **(2 marks)**

(b) Write the balanced ionic equation for the reaction described above.

.... **(2 marks)**

4 The following ionic equations are **not** balanced. Write the balanced equations in the spaces below them.

(a) $Fe^{2+} + OH^- \rightarrow Fe(OH)_2$

$Fe^{2+} + 2OH^- \rightarrow Fe(OH)_2$ **(1 mark)**

(b) $Fe^{3+} + OH^- \rightarrow Fe(OH)_3$

$Fe^{3+} + 3OH^- \rightarrow Fe(OH)_3$ **(1 mark)**

5 Alkaline solutions contain hydroxide ions. These react with hydrogen ions during neutralisation reactions.

(a) Write the ionic equation for the reaction between a hydrogen ion and a hydroxide ion.

.... **(1 mark)**

(b) Name the product of this reaction.

.... **(1 mark)**

6 When chlorine reacts with potassium bromide solution, potassium chloride solution and bromine form:

$Cl_2 + 2KBr \rightarrow 2KCl + Br_2$

> Potassium ions are **spectator ions** in this reaction. They are unchanged and can be left out of the equation.

(a) Write the formulae of all the ions present in this reaction.

.... **(3 marks)**

(b) Write a balanced ionic equation for the reaction.

.... **(2 marks)**

Hazards, risk and precautions

Guided

1   Describe what is meant by the term **hazard**.

A hazard is something that could cause *harm to To Somene or Somthing*

.... **(2 marks)**

2   Describe what is meant by the term **risk**.

> **Practical skills** Risk and hazard are **not** the same thing.

Arisk is the chance that somene or Something Will be harmed

(2 marks)

3   Hazard symbols are found on containers. Give **two** reasons why these hazard symbols are used.

Warn about the dangerour and precedion

.... **(2 marks)**

Guided

4   Complete the diagram below using a straight line to connect each hazard symbol to its correct description.

Symbol	Description
	**flammable** may easily catch fire
	**oxidising agent** may cause other substances to catch fire, or make a fire worse
	**corrosive** causes severe damage to skin and eyes
	**harmful or irritant** health hazard
	**toxic** may cause death by inhalation, ingestion or skin contact

(5 marks)

5   Copper reacts with concentrated nitric acid. The reaction forms copper nitrate, water and nitrogen dioxide. Nitrogen dioxide is a toxic brown gas with an irritating odour.

Explain a suitable precaution, other than eye protection, needed for safe working in this experiment.

Use gloves, use a different method

.... **(2 marks)**

Atomic structure

1 Which of these statements correctly describes an atom?

   ☑ **A** Most of the mass is concentrated in the nucleus.

   ☐ **B** Most of the charge is concentrated in the nucleus.

   ☐ **C** The number of neutrons always equals the number of protons.

   ☐ **D** The number of electrons always equals the number of neutrons. **(1 mark)**

2 Complete the table to show the relative mass, relative charge and position of each particle in an atom.

> Guided

Particle	proton	neutron	electron
Relative mass	+1	1	~~1~~ negligable
Relative charge	+1	0	−1
Position	nucleus	nucleas	aroun

(3 marks)

3 Explain why a hydrogen atom has no overall charge, even though it contains electrically charged particles.

....

.... **(2 marks)**

4 John Dalton described his atomic model of the atom in 1803. Suggest a reason that explains why his model did not include protons, neutrons and electrons.

.... They were not discovered then .... **(1 mark)**

5 The diameter of a gold atom is 2.70×10^{-10} m. The diameter of a gold nucleus is 1.03×10^{-14} m. Calculate, to three significant figures, the diameter of a gold atom relative to the diameter of its nucleus.

> 🖩 **Maths skills** 1.03×10^{-14} is written in standard form. You could enter it on your calculator as: 1.03 EXP −14.

.... 26200 .... **(2 marks)**

6 Experiments were carried out in the early part of the last century to test the 'plum pudding' model of the atom. A very large number of positively charged particles were fired at a very thin gold sheet.

   (a) Suggest a reason that explains why most of these particles passed straight through the gold sheet.

   .... Most of the nucleas is small .... **(1 mark)**

   (b) The positively charged particles are repelled when they come close to the nucleus of a gold atom. Explain what property of the nucleus is shown by this observation.

   .... The nucleas is positively charged ....
   .... So it repels the particle .... **(2 marks)**

   (c) In the experiments, only about 1 in 20 000 positively charged particles was repelled. Explain this observation.

   > You may be asked to analyse information and draw conclusions using your knowledge and understanding.

   .... The nucleas has a is very small compared ....
   .... to the cell .... **(2 marks)**

Isotopes

1 State what is meant by the **mass number** of an atom.

> Guided

The mass number of an atom is the total number of protons and neutrons in the nucleus

(1 mark)

2 An atom of an element X has an atomic number 9 and a mass number 19. How many electrons does an atom of element X contain?

☑ **A** 9   ☐ **B** 10   ☐ **C** 19   ☐ **D** 28   **(1 mark)**

3 Describe, in terms of particles in the atom, what an element is.

A element is a Substance that cannot be broken down into any other Substance as they have the Same number of photons

> What is the same for atoms of a given element, and what is different between atoms of different elements?

(2 marks)

4 Hydrogen has three natural isotopes: ^1_1H (hydrogen-1), ^2_1H (hydrogen-2) and ^3_1H (hydrogen-3).

(a) Complete the table to show the numbers of protons, neutrons and electrons in an atom of each isotope.

Isotope	Protons	Neutrons	Electrons
hydrogen-1	1	0	1
hydrogen-2	1	1	1
hydrogen-3	1	2	1

(3 marks)

(b) Explain, in terms of particles, why these are isotopes of the same element.

They are isotopes of the same element as they all have Same amount of electrons and

(2 marks)

5 Explain why relative atomic masses of some elements are whole numbers, but those of some other elements, for example chlorine, are not.

The relative atomic mass isn't the Mass Number

(2 marks)

6 A sample of neon contains two isotopes, ^{20}Ne (neon-20) and ^{22}Ne (neon-22).

> Guided

The relative abundance of neon-20 is 90.5%.

Calculate the relative atomic mass, A_r, of this sample of neon. Give your answer to one decimal place.

relative abundance of neon-22 = (100 – 90.5) = 9.5

mass of 100 atoms = 2606 2019

A_r of Ne = 20.19

(3 marks)

Mendeleev's table

1 (a) How did Mendeleev **first start** to arrange the elements in his periodic table?

☐ **A** in the order of increasing number of protons in the nucleus

☐ **B** in the order of increasing reactivity with other elements

☐ **C** in the order of increasing number of isotopes

☒ **D** in the order of increasing relative atomic mass **(1 mark)**

(b) State **one** factor, other than the one in your answer to part (**a**), that Mendeleev used when he arranged the elements.

.... properties   of thera elements with thier compand **(1 mark)**

2 The diagram shows part of Mendeleev's 1871 table.

(a) Give **two** similarities between this table and the modern periodic table.

> Remember that you will be given a periodic table in the exam. There is also one at the back of this book.

Group						
1	2	3	4	5	6	7
H						
Li	Be	B	C	N	O	F
Na	Mg	Al	Si	P	S	Cl
K Cu	Ca Zn	* *	Ti *	V As	Cr Se	Mn Br
Rb Ag	Sr Cd	Y In	Zr Sn	Nb Sb	Mo Te	* I

1 ...elements putinglaps

2 ...elements putin perra

(2 marks)

(b) Give **three** differences between this table and the modern periodic table.

1 ...There is 8 groups In mea

2 ...There is more than 1element in boa

3 ...had gaps **(3 marks)**

3 Mendeleev had difficulty placing some elements. For example, the order of tellurium Te and iodine I appeared to be reversed in his table.

(a) Explain why the positions of these two elements appeared to be reversed in Mendeleev's table.

.......He matched the properties y thee elemany........

....with there compromor................ ✗

(2 marks)

(b) Explain, in terms of atomic structure, why the positions of these two elements were actually correct.

....

....

.... **(2 marks)**

4 State **one** feature of Mendeleev's work with his table that would later help to support his ideas.

.... leaving gaps in the table..................... **(1 mark)**

The periodic table

1　How are the elements arranged in the modern periodic table?

☐　**A**　in the order of increasing mass number

☒　**B**　in the order of increasing atomic number

☐　**C**　in the order of increasing nucleon number

☐　**D**　in the order of increasing numbers of electron shells　**(1 mark)**

2　The positions of five elements (**A**, **B**, **C**, **D** and **E**) are shown in the periodic table on the right. These letters are **not** the chemical symbols for these elements.

(a)　Give the letters of **two** elements that have similar chemical properties to each other.

...................AB.................... **(1 mark)**

(b)　Give the letters of **all** the metal elements.

................A.B.C................ **(1 mark)**

(c)　Give the letters of **two** elements in the same period.

...........A B   BE............ **(1 mark)**

3　The meaning of the term **atomic number** has changed over time.

Guided

(a)　Explain the meaning of the term **atomic number** as Mendeleev might have understood it in the nineteenth century.

the position of ......the atomic number   in a reacuney...

.......relative mass........................ **(2 marks)**

(b)　Explain the modern meaning of the term **atomic number**.

.....The number of protons in the nucleus..... **(2 marks)**

(c)　Suggest a reason that explains why the meaning of atomic number has changed over time.

....In the 19th century the nucleus had.......

.... **(1 mark)**

4　Sodium is placed between elements **A** and **B** on the periodic table shown in question **2**. Argon is placed immediately above element **E**. Explain why there can only be six elements between sodium and argon.

> Think about why two different elements cannot occupy the same position on the modern periodic table.

There can only be 6 elements between ata A and B as Elments are arranged ra adr g increasing atomic number **(2 marks)**

Electronic configurations

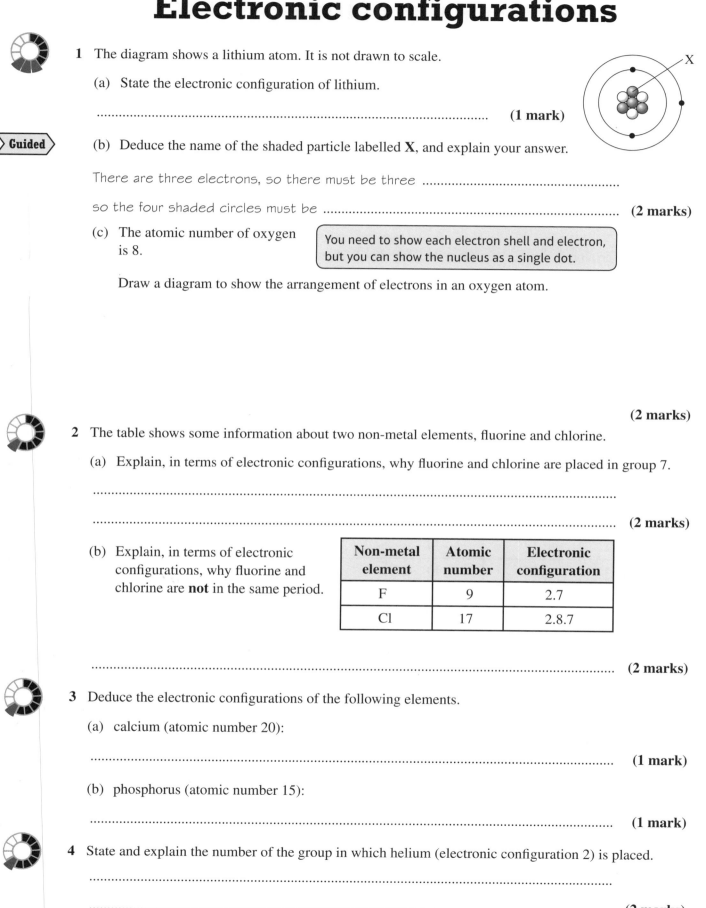

1 The diagram shows a lithium atom. It is not drawn to scale.

(a) State the electronic configuration of lithium.

.... **(1 mark)**

> **Guided**

(b) Deduce the name of the shaded particle labelled **X**, and explain your answer.

There are three electrons, so there must be three ....

so the four shaded circles must be .... **(2 marks)**

(c) The atomic number of oxygen is 8.

> You need to show each electron shell and electron, but you can show the nucleus as a single dot.

Draw a diagram to show the arrangement of electrons in an oxygen atom.

(2 marks)

2 The table shows some information about two non-metal elements, fluorine and chlorine.

(a) Explain, in terms of electronic configurations, why fluorine and chlorine are placed in group 7.

....

.... **(2 marks)**

(b) Explain, in terms of electronic configurations, why fluorine and chlorine are **not** in the same period.

Non-metal element	Atomic number	Electronic configuration
F	9	2.7
Cl	17	2.8.7

.... **(2 marks)**

3 Deduce the electronic configurations of the following elements.

(a) calcium (atomic number 20):

.... **(1 mark)**

(b) phosphorus (atomic number 15):

.... **(1 mark)**

4 State and explain the number of the group in which helium (electronic configuration 2) is placed.

....

.... **(2 marks)**

Ions

1 Which of the following statements correctly describes the formation of an ion?

☐ **A** Positively charged ions, called cations, form when atoms or groups of atoms gain electrons.

> You can quickly narrow the alternatives if you know the correct name for each type of ion, or how it forms.

☐ **B** Positively charged ions, called anions, form when atoms or groups of atoms lose electrons.

☐ **C** Negatively charged ions, called cations, form when atoms or groups of atoms lose electrons.

☐ **D** Negatively charged ions, called anions, form when atoms or groups of atoms gain electrons.

(1 mark)

2 The atomic number of magnesium, Mg, is 12. The symbol for a magnesium ion is Mg^{2+}.

(a) Deduce the number of electrons in a magnesium ion.

> **Maths skills** Work out the number of electrons in an atom, then add or subtract electrons according to the charge shown.

.... **(1 mark)**

(b) Write the electronic configuration for a magnesium ion.

.... **(1 mark)**

3 Complete the table to show the numbers of protons, neutrons and electrons in each ion.

Guided

Ion	Atomic number	Mass number	Protons	Neutrons	Electrons
N^{3-}	7	15	7	8	10
K^+	19	40			
Ca^{2+}	20	40			
S^{2-}	16	32			
Br^-	35	81			

(5 marks)

4 The diagram on the right shows the formation of a sodium ion, Na^+, from a sodium atom.

Draw a similar diagram to show the formation of a chloride ion, Cl^-, from a chlorine atom.

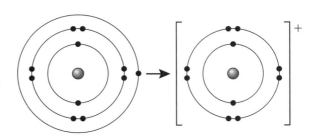

(3 marks)

Formulae of ionic compounds

1   The formula of a sodium ion is Na^+. The formula of a phosphate ion is PO_4^{3-}. Which of the following is the formula for sodium phosphate?

☐ **A** $NaPO_4$

☐ **B** $Na(PO_4)_3$

☐ **C** Na_2PO_4

☐ **D** Na_3PO_4 **(1 mark)**

2   Complete the table to show the formulae of the compounds produced by each pair of ions.

> **Guided**

> 🖩 **Maths skills** You may need more than one of each ion to obtain equal numbers of positive and negative charges.

> You need to know the formulae of common ions. This helps you work out the formulae of ionic substances.

	Cl^-	S^{2-}	OH^-	NO_3^-	SO_4^{2-}
K^+				KNO_3	
Ca^{2+}			$Ca(OH)_2$		$CaSO_4$
Fe^{3+}		Fe_2S_3			
NH_4^+	NH_4Cl				

(20 marks)

3   Magnesium ribbon burns in air. It reacts with oxygen to produce magnesium oxide, MgO.

(a)   Write the balanced equation for the reaction.

.... **(2 marks)**

(b)   Magnesium nitride is also formed, as some of the hot magnesium reacts with nitrogen in the air.

(i)   Nitrogen is in group 5. Suggest reasons that explain why the formula for a nitride ion is N^{3-}.

....

.... **(2 marks)**

(ii)   Write the formula for magnesium nitride.

> The formula for a magnesium ion is Mg^{2+}.

.... **(1 mark)**

(iii) Explain why the NO_3^- ion is called the nitrate ion, but the N^{3-} ion is called the nitride ion.

....

.... **(2 marks)**

4   Complete the table to show the names of the ions.

> Remember to use the endings -ide and -ate correctly.

	S^{2-}	SO_4^{2-}	Cl^-	ClO_3^-
Name				

(4 marks)

Properties of ionic compounds

1 Which statement about the formation of ionic compounds, such as sodium chloride, is correct?

☐ **A** Electrons are transferred from metal atoms to non-metal atoms, producing cations and anions.

☐ **B** Electrons are transferred from cations to anions, producing metal atoms and non-metal atoms.

☐ **C** Electrons are shared between metal atoms and non-metal atoms.

☐ **D** Electrons are shared between cations and anions.   **(1 mark)**

2 Ionic compounds have a lattice structure.

(a) Complete the diagram, using the symbols + and −, to show the positions of positive and negative ions in an ionic lattice.

> **Maths skills** Remember that opposite charges will attract each other, and like charges will repel.

> You should be able to visualise and represent 2D and 3D forms, including 2D representations of 3D objects.

(1 mark)

(b) Describe what ionic bonds are.

....

.... **(2 marks)**

3 (a) Explain why ionic compounds have high boiling points.

> Mention the forces between the particles found in ionic compounds.

....

.... **(2 marks)**

(b) Suggest a reason that explains why the melting point of MgO is higher than the melting point of NaCl.

.... **(1 mark)**

4 Calcium metal can be produced on an industrial scale by passing an electric current through molten calcium chloride.

> **Guided**

(a) Explain why molten calcium chloride can conduct electricity.

When calcium chloride is a liquid, its ions are ....

.... **(2 marks)**

(b) State why solid calcium chloride **cannot** conduct electricity.

.... **(1 mark)**

(c) Describe one way, other than by melting it, of making calcium chloride conduct electricity.

.... **(1 mark)**

Covalent bonds

1 What are the typical sizes of atoms and small molecules?

> **Maths skills** The quantities are shown in standard form. For example, 10^{-3} is greater than 10^{-6}.

	☐ A	☐ B	☐ C	☐ D
Atoms	10^{-10} m	10^{-10} m	10^{-9} m	10^{-12} m
Molecules	10^{-11} m	10^{-9} m	10^{-12} m	10^{-9} m

(1 mark)

2 Explain how a covalent bond forms.

....

.... **(2 marks)**

3 Hydrogen reacts with fluorine to form hydrogen fluoride: $H_2 + F_2 \rightarrow 2HF$

The electronic configuration of hydrogen is 1 and the electronic configuration of fluorine is 2.7.

> **Guided**

(a) Explain why fluorine atoms can form only one covalent bond.

A fluorine atom has one unpaired electron in its ....

so it .... **(2 marks)**

(b) Describe what the structure, H–H, tells you about a hydrogen molecule.

....

.... **(2 marks)**

(c) Draw the dot-and-cross diagrams for a molecule of each of the following substances, showing the outer electrons only.

> Show each chemical symbol. Show one atom's electrons as dots and the other atom's electrons as crosses.

    (i) fluorine:

(2 marks)

    (ii) hydrogen fluoride:

(2 marks)

4 The electronic configuration of nitrogen is 2.5.

(a) Draw a dot-and-cross diagram for a nitrogen molecule, N_2. Show the outer electrons only.

(2 marks)

(b) Draw the structure for a nitrogen molecule.

> Look at question **3b**.

.... **(1 mark)**

Simple molecular substances

1 Carbon dioxide, CO_2, is found in the air. Why does it have a low boiling point?

☐ **A** There are weak forces of attraction between carbon atoms and oxygen atoms.

☐ **B** There are weak covalent bonds between carbon atoms and oxygen atoms.

☐ **C** There are weak forces of attraction between carbon dioxide molecules.

☐ **D** There are weak covalent bonds between carbon dioxide molecules.   **(1 mark)**

2 The table shows the properties of three different substances (**A**, **B** and **C**).

Substance	Melting point (°C)	Solubility in water (g per 100 g of water)	Conducts electricity when solid?	Conducts electricity when liquid?
A	290	43	no	yes
B	−95	0.001	no	no
C	660	0	yes	yes

State and explain which substance (**A**, **B** or **C**) is a simple molecular substance.

....

....

.... **(3 marks)**

3 Sulfur hexafluoride, SF_6, exists as simple molecules. It is used as an insulating gas for electrical equipment.

(a) Explain why sulfur hexafluoride does not conduct electricity.

> Think about whether simple molecules are electrically charged or contain electrons that are free to move.

....

.... **(2 marks)**

> **Guided**

(b) Suggest reasons that explain why sulfur hexafluoride does not dissolve in water.

The intermolecular forces between ....

are weaker than those between ....

and those between... **(3 marks)**

4 The graph shows the boiling points of three alcohols. Their relative formula masses are shown on each bar.

Describe the relationship shown by the graph, and suggest a reason that explains it.

....

....

.... **(2 marks)**

Giant molecular substances

1 Silica, SiO_2, does not conduct electricity or dissolve in water. Its melting point is very high.

Which statement describes a molecule of silica?

☐ **A**   a giant molecule with ionic bonds   ☐ **C**   a simple molecule with covalent bonds

☐ **B**   a giant molecule with covalent bonds   ☐ **D**   a simple molecule with ionic bonds

(1 mark)

2 The diagrams below show the structures of diamond and graphite.

> You should be able to visualise and represent 2D and 3D forms, including 2D representations of 3D objects.

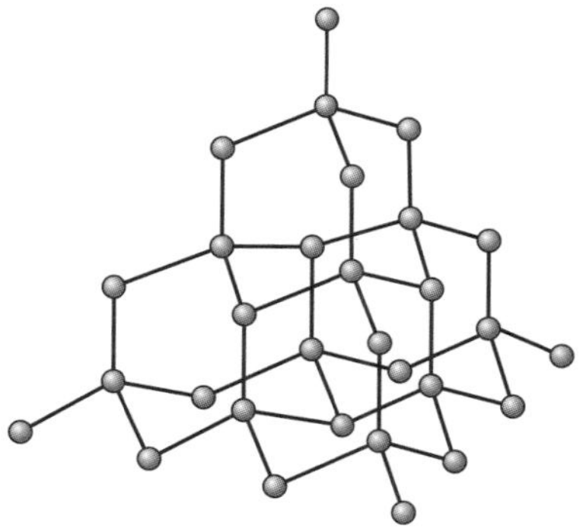

diamond 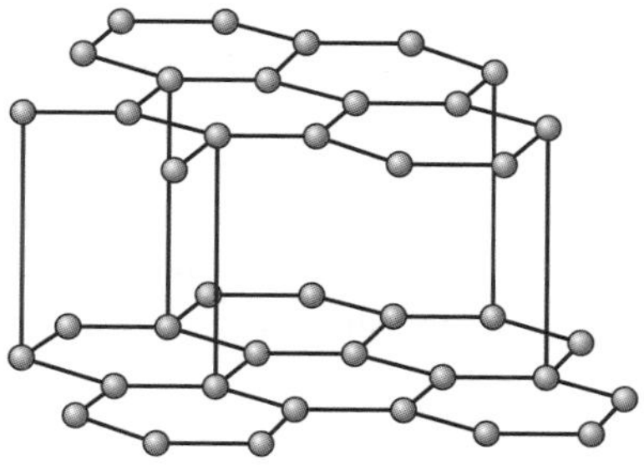   graphite

(a) Name the element that has atoms represented by the balls in the diagrams.

.... **(1 mark)**

(b) State the maximum number of bonds present between each atom in a molecule of diamond.

.... **(1 mark)**

(c) Name the type of structure shown in both diagrams.

.... **(1 mark)**

3 Refer to structure and bonding in your answers to the following questions.

> You need to explain why diamond is very hard.

(a) Explain why diamond is suitable for use in cutting tools.

....

.... **(3 marks)**

Guided

(b) Explain why graphite is suitable for use as a lubricant.

The layers in graphite can ....

because .... **(2 marks)**

(c) Explain why graphite is used to make electrodes.

> You need to explain why graphite can conduct electricity.

....

....

.... **(2 marks)**

Other large molecules

1 Ethene, C_2H_4, can be made into a polymer. What is the name of this polymer?

☐ **A** plastic

☐ **B** poly(ethane)

☐ **C** poly(ethene)

☐ **D** poly(ethyne) **(1 mark)**

2 The diagram is a model of a section of a simple polymer.

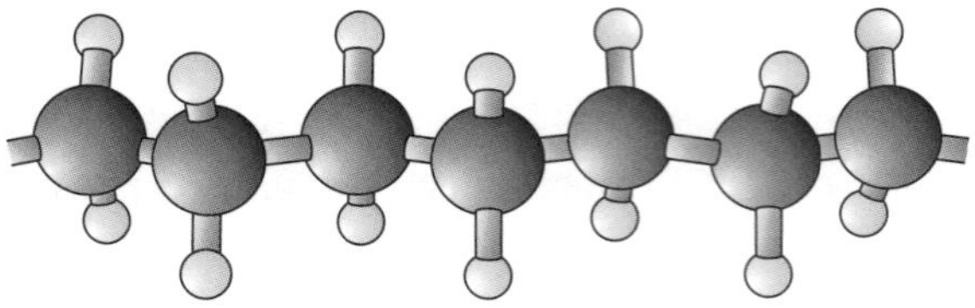

(a) Name the element with atoms represented by the larger, dark-grey balls in the diagram.

.... **(1 mark)**

(b) Name the type of bonding present in a molecule of this polymer.

.... **(1 mark)**

3 Graphene is a form of carbon. It is a good conductor of electricity and has a very high melting point.

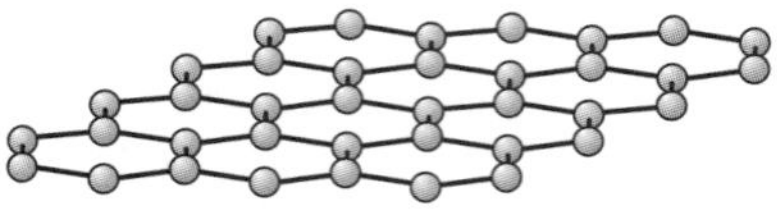

The diagram is a model of part of the structure of graphene.

▷ **Guided**

(a) Explain, in terms of its structure and bonding, why graphene has a very high melting point.

> Include the type of bonds that must be broken during melting.

Graphene has .... bonds in a ....

structure, and these bonds are.. **(3 marks)**

(b) Explain why graphene is a good conductor of electricity.

> Graphene has a structure similar to a layer of graphite.

....

....

.... **(2 marks)**

4 Fullerenes are forms of carbon that include hollow balls, such as buckminsterfullerene, C_{60}.

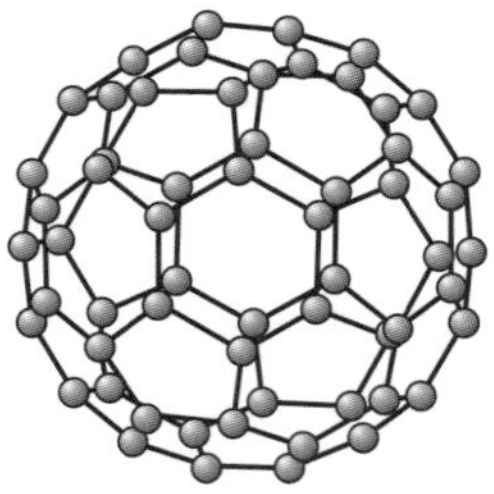

Explain, in terms of bonding, why buckminsterfullerene has a much lower melting point than graphite.

....

....

.... **(3 marks)**

Metals

1   Metal elements and non-metal elements have different typical properties.

Complete the table below by placing a tick (✓) in each correct box.

	Low melting points	High melting points	Good conductors of electricity	Poor conductors of electricity
Metals				
Non-metals				

(4 marks)

2   Most metals are shiny solids with high densities. Explain what having a 'high density' means.

....

.... **(2 marks)**

3   Copper is a metal used in electricity cables. It is a good conductor of electricity and is malleable (it will bend without shattering). The diagram is a model for the structure of copper. Each circle is a copper ion.

(a)   State two improvements to the diagram that will make it a more accurate model of the structure of copper.

> Remember that ions are charged particles.

....

.... **(2 marks)**

> **Guided**

(b)   Explain why copper is malleable.

Layers of ....

can .... **(2 marks)**

(c)   Explain why copper is a good conductor of electricity.

....

.... **(2 marks)**

4   Explain why many metals have high melting points, using ideas about metallic bonding to justify your answer.

> In your answer, mention which particles are attracted to each other in a metal crystal.

....

....

....

.... **(3 marks)**

5   Metals are insoluble in water. However, when a granule of calcium is added to water, it fizzes and gradually disappears. Suggest an explanation for these observations.

....

.... **(2 marks)**

Limitations of models

1 The formula of a substance can be given in different ways.

Which row (**A**, **B**, **C** or **D**) correctly shows the different formulae for ethene?

	Molecular formula	Empirical formula	Structural formula
☐ A	C_2H_6	CH_3	CH_3CH_3
☐ B	C_2H_4	CH_2	$CH_2=CH_2$
☐ C	CH_2	C_2H_4	$CH_2=CH_2$
☐ D	$CH_2=CH_2$	C_2H_4	CH_2

> Answer **A** cannot be correct because it describes ethane, not ethene.

(1 mark)

2 The diagrams (**A**, **B**, **C** and **D**) show four different models for a molecule of methane, CH_4.

A	B	C	D
$H-\overset{\displaystyle H}{\underset{\displaystyle H}{C}}-H$			
Structure	**Dot-and-cross diagram**	**Ball-and-stick model**	**Space-filling model**

State the letters (**A**, **B**, **C** or **D**) for the models that:

> You may need to identify more than one model in your answers.

(a) show the covalent bonds present in a methane molecule

.... **(1 mark)**

(b) identify the elements present in a methane molecule

.... **(1 mark)**

(c) represent the three-dimensional shape of a methane molecule

.... **(1 mark)**

(d) show the electrons involved in bonding

.... **(1 mark)**

(e) show the relative sizes of each atom in a methane molecule

.... **(1 mark)**

3 A student draws a dot-and-cross diagram of a water molecule. Compare and contrast the advantages and disadvantages of drawing a ball-and-stick model instead.

> Think about the limitations of each model. You do not need to write a conclusion in your answer.

....

....

....

.... **(3 marks)**

Relative formula mass

Use the relative atomic masses, A_r, in the table below when you answer the questions.

Element	Al	Ca	Cl	Cu	H	N	O	S
A_r	27	40	35.5	63.5	1	14	16	32

1 Calculate the relative formula mass, M_r, of each of the following substances.

> You do not need to show your working out, but it will help you to check the accuracy of your answers.

> If relative atomic masses are not given in the question, you can find them in the periodic table.

(a) water, H_2O

........18........ **(1 mark)**

(b) sulfur dioxide, SO_2

........64........ **(1 mark)**

(c) aluminium oxide, Al_2O_3

........102........ **(1 mark)**

(d) ammonium chloride, NH_4Cl > Do not round the answer to this question to a whole number.

........53.5........ **(1 mark)**

(e) calcium chloride, $CaCl_2$

........111........ **(1 mark)**

(f) aluminium chloride, $AlCl_3$

........133.5........ **(1 mark)**

2 Calculate the relative formula mass, M_r, of each of the following substances.

> Guided

(a) calcium hydroxide, $Ca(OH)_2$ 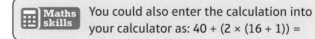 **Maths skills** You could also enter the calculation into your calculator as: $40 + (2 \times (16 + 1)) =$

16 + 1 = 17, 17 × 2 = 34, 40 + 34 = ...74........ **(1 mark)**

(b) aluminium hydroxide, $Al(OH)_3$

........78........ **(1 mark)**

(c) calcium nitrate, $Ca(NO_3)_2$

........164........ **(1 mark)**

(d) ammonium sulfate, $(NH_4)_2SO_4$

.... **(1 mark)**

(e) aluminium sulfate, $Al_2(SO_4)_3$

.... **(1 mark)**

 Empirical formulae

1   A student carries out an experiment to determine the empirical formula of magnesium oxide. He heats a piece of magnesium ribbon in a crucible. He continues until the contents of the crucible stop glowing.

The table shows his results.

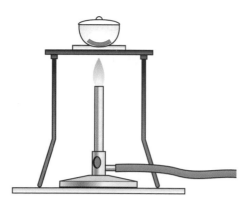

Object	Mass (g)
empty crucible and lid	20.24
crucible, lid and contents before heating	20.49
crucible, lid and contents after heating	20.65

(a) Suggest a reason that explains why the student continued heating until the contents stopped glowing.

.... **(1 mark)**

(b) The hot crucible is a hazard. Explain one precaution needed to control the risk of harm.

....

.... **(2 marks)**

> **Guided** >   (c) Calculate the empirical formula of magnesium oxide using the student's results.

(A_r of Mg = 24 and A_r of O = 16)

mass of magnesium used = 20.49 g – 20.24 g = 0.25 g

mass of oxygen reacted = 20.65 g – 20.49 g = ...

Mg   **O**

$\frac{0.25}{24} = 0.0104$ $\frac{.......}{16} = ...$

> Divide the mass of each element by its A_r.

> Divide both numbers by the smallest number to find the ratio.

$\frac{0.0104}{.......} = ...$ $\frac{.............}{0.010} = ...$

Empirical formula is ....

> Write down the empirical formula.

(5 marks)

2   In an experiment, 11.2 g of hot iron reacts with 21.3 g of chlorine gas to form iron chloride.

Calculate the empirical formula of the iron chloride.

(A_r of Fe = 56 and A_r of Cl = 35.5)

.... **(3 marks)**

3   The empirical formula of a sample of gas is NO_2. Its relative formula mass, M_r, is 92.

Deduce the molecular formula of the gas.

....

.... **(2 marks)**

Conservation of mass

1 Sodium chloride solution reacts with silver nitrate solution. Sodium nitrate solution and a white precipitate of solid silver chloride form: $NaCl(aq) + AgNO_3(aq) \rightarrow NaNO_3(aq) + AgCl(s)$

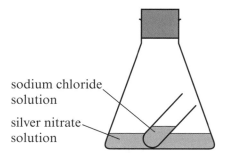

sodium chloride solution

silver nitrate solution

A student investigates the change in mass during this reaction. He sets up the apparatus shown in the diagram. He finds the total mass of the flask and its contents, and then shakes the flask to mix the solutions.

(a) Explain whether the reaction takes place in a closed or non-enclosed system.

....

.... **(1 mark)**

(b) Explain what will happen to the total mass of the flask and its contents during the reaction.

....

.... **(2 marks)**

2 Sodium reacts with chlorine to form sodium chloride: $2Na(s) + Cl_2(g) \rightarrow 2NaCl(s)$

Guided

Calculate the maximum mass of sodium chloride that can be made from 21.3 g of chlorine.

(M_r of Cl_2 = 71 and M_r of NaCl = 58.5)

$(1 \times 71) = 71$ g of Cl_2 makes $(2 \times 58.5) = 117$ g of NaCl

21.3 g of Cl_2 makes $117 \times \left(\dfrac{21.3}{71}\right)$ g of NaCl

= .... g   **(2 marks)**

3 Magnesium reacts with oxygen to form magnesium oxide: $2Mg(s) + O_2(g) \rightarrow 2MgO(s)$

> Remember to calculate the relative formula mass, M_r, of oxygen gas and magnesium oxide first.

Calculate the maximum mass of magnesium oxide that can be made from 12.6 g of oxygen.

(A_r of O = 16 and A_r of Mg = 24)

.... **(3 marks)**

4 Calcium carbonate decomposes, when heated, to form calcium oxide and carbon dioxide:

$CaCO_3(s) \rightarrow CaO(s) + CO_2(g)$

Calculate the maximum mass of calcium oxide that can be made from 12.5 kg of calcium carbonate.

(A_r of Ca = 40, A_r of C = 12 and A_r of O = 16)

.... **(3 marks)**

Reacting mass calculations

1 Magnesium ribbon reacts with dilute hydrochloric acid. Magnesium chloride solution and hydrogen gas form: $Mg(s) + 2HCl(aq) \rightarrow MgCl_2(aq) + H_2(g)$

Which of the following statements about this reaction is correct?

> A **limiting reactant** is a reactant that is not in excess.

☐ **A** When magnesium is in excess, no magnesium is left when the reaction stops.

☐ **B** When hydrochloric acid is the limiting reactant, no liquid is left when the reaction stops.

☐ **C** When hydrochloric acid is in excess, some magnesium is left when the reaction stops.

☐ **D** When hydrochloric acid is the limiting reactant, some magnesium is left when the reaction stops.

(1 mark)

2 Copper carbonate decomposes when heated, forming copper oxide and carbon dioxide gas:

$$CuCO_3(s) \rightarrow CuO(s) + CO_2(g)$$

The graph shows how the mass of copper oxide formed depends on the mass of copper carbonate heated.

(a) Describe the relationship shown by the graph.

> **Maths skills** The mass of copper oxide formed is the **dependent variable** and the mass of copper carbonate heated is the **independent variable**. What happens to the dependent variable as the independent variable changes?

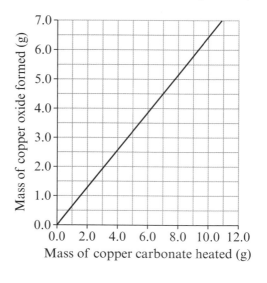

....

.... **(2 marks)**

(b) Explain why the mass of carbon dioxide formed depends on the mass of copper carbonate used.

....

.... **(2 marks)**

Guided

3 Iron is heated with excess chlorine gas, Cl_2, forming iron(III) chloride, $FeCl_3$.

11.2 g of iron produces 32.5 g of iron(III) chloride. Use this information to determine the stoichiometry of the reaction. (A_r of Fe = 56, M_r of Cl_2 = 71 and M_r of $FeCl_3$ = 162.5)

mass of chlorine reacted = 32.5 g – 11.2 g = ... g

amount of iron = $\dfrac{11.2}{56}$ = ... mol

amount of chlorine = $\dfrac{............}{71}$ = ... mol

amount of iron(III) chloride = $\dfrac{32.5}{162.5}$ = ... mol

ratio of Fe : Cl_2 : $FeCl_3$ = ...:.......:.......

whole number ratio of Fe : Cl_2 : $FeCl_3$ = ...:......:......

> Divide each number by the smallest number, then multiply all by the same whole number if necessary.

so equation must be ....

(4 marks)

Concentration of solution

1 Calculate the following volumes in dm³.

(a)　2000 cm³

.... **(1 mark)**

(b)　500 cm³

.... **(1 mark)**

(c)　25 cm³

.... **(1 mark)**

2 A student dissolves 10 g of copper sulfate in 250 cm³ of water. Calculate the concentration of the solution formed in g dm⁻³.

Guided

$\text{concentration} = \left(\dfrac{10}{250}\right) \times 1000 = $ .... **(1 mark)**

3 Calculate the concentrations of the following solutions in g dm⁻³.

(a)　5.0 g of sodium hydroxide dissolved in 100 cm³ of water

.... **(1 mark)**

(b)　14.6 g of hydrogen chloride dissolved in 400 cm³ of water

.... **(1 mark)**

(c)　0.25 g of glucose dissolved in 25 cm³ of water

.... **(1 mark)**

4 A school technician wants to make 2.5 dm³ of a 40 g dm⁻³ aqueous solution of sodium hydroxide.

(a)　Describe the meaning of the term **aqueous solution**.

.... **(1 mark)**

(b)　Calculate the mass of sodium hydroxide that the technician must dissolve to make her solution.

.... **(1 mark)**

Guided

(c)　A student mixes 50 cm³ of the technician's solution with 200 cm³ of water.

Calculate the concentration of the student's sodium hydroxide solution in g dm⁻³.

Mass of NaOH in 50 cm³ $= 40 \times \dfrac{50}{1000} = $ .... g

New concentration $= \left(\dfrac{\text{mass}}{250}\right) \times 1000 = $ .... **(2 marks)**

Avogadro's constant and moles

1 One mole, 1 mol, of particles of a substance contains 6.02×10^{23} particles.

 Maths skills 6.02×10^{23} is a number expressed in standard form.
You can enter it into your calculator as: 6.02 EXP 23.

State the name given to this number.

.... Avogrado's constant .... **(1 mark)**

2 Calculate the mass of the following substances.

The relative atomic mass, A_r, of carbon is 12. This means that 1 mol of carbon atoms has a mass of 12 g.

(a) 0.500 mol of xenon atoms, Xe (A_r of Xe = 131)

.... 65.5 g .... **(1 mark)**

(b) 1.5 mol of oxygen molecules, O_2 (A_r of O = 16)

.... 2498 g .... **(1 mark)**

(c) 0.30 mol of ammonium ions, NH_4^+
(A_r of N = 14 and A_r of H = 1)

.... 5.4 g ✓ .... **(1 mark)**

3 Calculate the amount, in mol, of the following substances.

Guided

(a) 6 g of carbon atoms, C (A_r of C = 12)

amount $= \dfrac{6}{12} = $ .... 6 ...... mol   **(1 mark)**

(b) 45 g of water molecules, H_2O (A_r of H = 1 and A_r of O = 16)

.... 2.5 .... **(1 mark)**

(c) 15 g of carbonate ions, CO_3^{2-} (A_r of C = 12 and A_r of O = 16)

.... 0.25 .... **(1 mark)**

4 Calculate the amount of **atoms**, in mol, in the following substances.

(a) 1.0 mol of carbon dioxide molecules, CO_2

.... 96.3 .... **(1 mark)**

(b) 0.5 mol of ethanoic acid molecules, CH_3COOH

.... 30 gas ⁷ .... **(1 mark)**

(c) 21.3 g of chlorine gas, Cl_2 (A_r of Cl = 35.5)

$35.5 \times 2 = 71 g,$ $21.3 \div 71 = 0.3$ g.mol........... 7.565 0.6 mol **(1 mark)**
$\div 2 = 0.6$ mol

5 Calculate the number of molecules in 2.25 mol of carbon dioxide molecules, CO_2.

$6.02 \times 10^{23} \times 2.25 = 1.35$ .... 1.35 .... **(1 mark)**

6 Calculate the amount of molecules, in mol, in 9.03×10^{24} molecules of water, H_2O.

.... 15 mol .... **(1 mark)**

mol × av cons = molecules

Extended response – Types of substance

Graphite and diamond are two different forms of carbon. The table shows some information about their properties. Copper, a soft metal used in electrical cables, is included for comparison.

Substance	Relative hardness	Relative electrical conductivity
graphite	10	10^8
copper	100	10^{10}
diamond	10 000	1

Higher values mean harder or better at conducting electricity.

Graphite is used to make electrodes and as a lubricant. Diamond is used in cutting tools. Explain each use in terms of the bonding and structure present. You should use information from the table in your answer.

You should be able to describe the structures of graphite and diamond.

You will be more successful in extended response questions if you plan your answer before you start writing. For example, write separate answers about graphite and diamond. Include each given use and, using information from the table, the property important to that use. Make sure that you then explain how the substance's bonding and structure give it that property.

Quick, labelled diagrams showing the structures of diamond and graphite may help your explanations.

....

....

....

....

....

....

....

....

....

....

....

....

.... **(6 marks)**

Had a go ☐ Nearly there ☐ Nailed it! ☐

States of matter

1 Iodine crystals become a purple vapour when they are warmed. What is the name for this state change?

> The crystals are in the solid state and the vapour is in the gas state.

☐ **A** melting   ☒ **C** subliming

☐ **B** boiling   ☐ **D** condensing   **(1 mark)** ①

2 Most substances can exist in the solid, liquid or gas states. Give the name of each state change below.

(a) liquid to solid

..........freezing... **(1 mark)** ①

(b) gas to liquid

..............Condensation... **(1 mark)** ②①

3 Water changes to steam when it is heated. State why this is a **physical** change.

.....It...can...be....easily...reversible.............................. **(1 mark)** ①

4 (a) Describe the arrangement, and movement, of particles in each state of matter.

> Is the arrangement regular or random? Are the particles close or far apart? How do the particles move?

solid ...Are...compacted...together...the...particles...don't...... ②
....move only...vibrate..

liquid ...In no...specific...order...and...can...more around...... ②
....but aren't...fully free...

gas ...fully...free...has...the...most energy...and........... ②
....move...the furthest................................ **(6 marks)**

(b) Particles in all states have some stored energy.

Name the state in which the particles have the most stored energy, and justify your answer.

..........Gas...as it moves the most.............................. ②
.... **(2 marks)**

5 Describe what happens to the arrangement, and movement, of particles when a substance changes from the liquid state to the solid state.

> Guided

The arrangement changes from......Solid...to...liquid............................

and the movement changes from...Particles...have...more energy........... ①
....and...more...around....in behing...each other................ **(2 marks)**

6 The melting point of substance **X** is $-114\,°C$ and its boiling point is $78\,°C$. Predict its state at $-30\,°C$. ①

....................liquid.. **(1 mark)**

Pure substances and mixtures

1 (a) Explain why sodium, Na, and chlorine, Cl_2, are two different elements.

> **Guided**

The atoms of an element all have the same ....

but atoms of Na and Cl_2 have different .... **(2 marks)**

(b) Explain why sodium chloride, NaCl, is defined as a compound.

...It is a compound because there are 2 or more...

...elements chemically Joined together... **(2 marks)**

2 A student investigates three samples of water. She transfers 25 cm³ of each sample to weighed evaporating basins. She then heats the basins until all the water has evaporated, lets them cool and weighs them again.

The table shows the student's results.

Water sample	Mass of basin before adding water (g)	Mass of basin after evaporating water (g)	Difference in mass (g)
A	73.05	73.20	0.15
B	72.61	72.85	0.25
C	74.40	74.43	0.03

> **Guided**

(a) Complete the table to show the difference in mass for water samples **B** and **C**. **(1 mark)**

(b) Explain whether any of the water samples are pure.

...No as if it was only pure water there would...

...be no difference in mass... **(2 marks)**

3 Solders are alloys used to join copper pipes or electrical components together. Some 'lead-free' solders are mixtures of tin and silver. The table shows the melting points of tin, silver and a lead-free solder.

Explain how the data show that the solder is a mixture.

> Look at how the melting points are shown for each substance.

Substance	Melting point (°C)
tin	232
silver	962
solder	220–229

....

....

.... **(2 marks)**

4 The boiling point of pure water is 100.0 °C. A sample of seawater boils at 100.6 °C.

(a) Seawater contains dissolved salts. State the effect of these salts on the boiling point of water.

.... **(1 mark)**

(b) A sample of seawater is boiled in an open container. Predict what will happen to the boiling point of the sample during this process, and justify your answer.

....

....

.... **(3 marks)**

Distillation

1   Which of the following is a suitable method to separate a mixture of two miscible liquids?

> Miscible liquids mix completely with each other.

☐ **A** filtration

☐ **B** simple distillation

☒ **C** fractional distillation

☐ **D** paper chromatography   **(1 mark)**

2   The apparatus shown in the diagram is used to separate the components of a seawater sample.

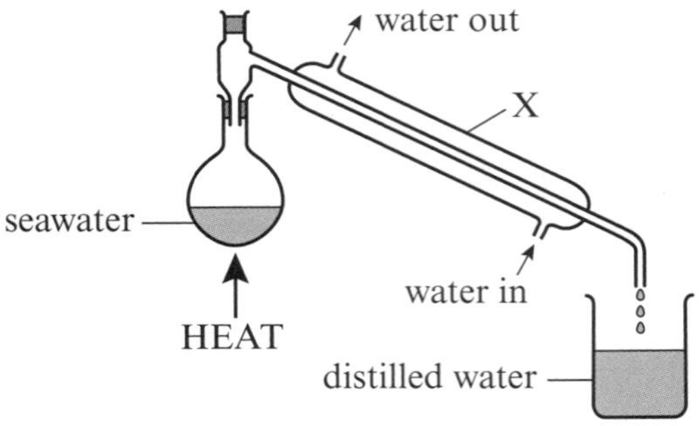

labels: water out, X, seawater, water in, HEAT, distilled water

(a) Describe what happens to vapour from the seawater as it passes through the apparatus labelled **X**.

The cold water runs through making keeping the conduser cold   **(2 marks)**

> **Guided**

(b) Cold water passes through the condenser. Explain what happens to its temperature.

The temperature of the water ...is hotter is cool... because cold water runs through   **(2 marks)**

3   The apparatus shown in the diagram is used to separate a mixture of ethanol and water.

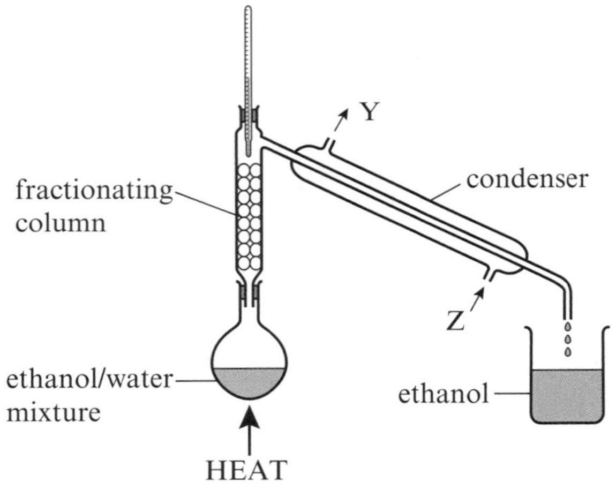

labels: Y, fractionating column, condenser, ethanol/water mixture, Z, ethanol, HEAT

(a) Give a reason that explains why ethanol and water can be separated using the method shown.

The liquids have a different boiling point   **(1 mark)**

(b) Explain which liquid, ethanol or water, is collected first.

Ethal water has ethanol boiling point is lower   **(2 marks)**

(c) Give a reason that explains why the cold water supply should be connected at **Z** rather than **Y**.

The the condser hols cold and the vapour stays the vapor doesn't pass to the condenser   **(1 mark)**

Filtration and crystallisation

1 A student filters a mixture of sand, salt and water. He collects the liquid that passes through the filter paper. Complete the table by placing a tick (✓) in the box against each correct statement.

Statement	Tick (✓)
The liquid is water.	
The liquid is the filtrate.	
The salt is left behind as a residue.	
The sand is left behind as a residue.	

> Do not place a cross against the incorrect statement(s) – you are asked to place only ticks in the table.

(2 marks)

2 Potassium iodide solution reacts with lead nitrate solution. A mixture of potassium nitrate solution and insoluble yellow lead iodide forms. When this is filtered, lead iodide remains in the filter paper.

> **Guided**

  (a) Balance the equation below, and give the state symbols for each substance.

    ......KI(aq) + $Pb(NO_3)_2$(......) → ......KNO_3(......) + PbI_2(......)   **(2 marks)**

  (b) The yellow lead iodide is washed, with distilled water, while it is on the filter paper.

    (i) State why the lead iodide does not pass through the filter paper.

    .... **(1 mark)**

    (ii) Suggest a reason that explains why the lead iodide is washed.

    .... **(1 mark)**

3 A student decides to make pure, dry copper chloride crystals. She adds an excess of insoluble copper carbonate to dilute hydrochloric acid. Copper chloride solution forms.

  (a) Name a suitable method to remove excess copper carbonate from the copper chloride solution.

  .... **(1 mark)**

  (b) The student leaves her copper chloride solution on a windowsill.

    (i) Explain why crystals of copper chloride form after a few days.

    ....

    ....

    .... **(3 marks)**

    (ii) Give a reason that explains why the student pours away the remaining solution, and then pats the crystals with filter paper.

    .... **(1 mark)**

  (c) The student could heat her copper chloride solution instead of leaving it on a windowsill. Describe two steps that she should take to obtain large, regular-shaped crystals.

  ....

  ....

  .... **(2 marks)**

Paper chromatography

1 Paper chromatography is used to determine whether an orange squash drink, **O**, contains an illegal food colouring, **X**.

Spots of each substance, and spots of three legal food colourings (**A**, **B** and **C**), are added to chromatography paper. The diagram shows the result of the chromatography experiment.

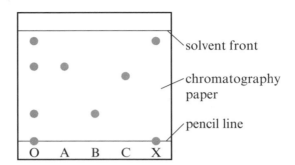

(a) Suggest a reason that explains why the start line is drawn using a pencil, rather than using ink.

.........*There is ink inside the pen line*......................... ✗ **(1 mark)**

(b) Explain whether the orange squash, **O**, is a pure substance or a mixture.

.........*The orange squash is a mixture as it stops at different*.........

.........*parts of the paper*......................... **(2 marks)**

(c) Identify which legal food colourings (**A**, **B** and/or **C**) are present in the orange squash.

.........*A, B*......................... **(1 mark)**

(d) Explain whether the orange squash contains the illegal food colouring, **X**.

.........*The orange squash contains the illegal food colouring as*.........

.........*both of the dots are in the same place for both O and X*... **(2 marks)**

(e) Suggest a reason that explains why one of the substances in **X** remains on the pencil line.

.........*Its less soluble*......................... **(1 mark)**

2 The diagram shows a chromatogram of a dye.

Calculate the R_f value of the spot in the chromatogram.

$$R_f = \frac{\text{distance travelled by a spot}}{\text{distance travelled by the solvent front}}$$

$$\frac{3}{3.5} = 0.8mm$$

.... **(3 marks)**

 **Practical skills**

Investigating inks

1 Paper chromatography is used to separate mixtures of coloured substances, such as those in inks.

(a) State two measurements that must be made so that an R_f value can be calculated.

...... distance travelled by the spot, Distance travelled by the Solvent ...... **(2 marks)**

(b) Suggest reasons that explain why these measurements are recorded to the nearest millimetre, rather than to the nearest centimetre.

> 🖩 **Maths skills** You must be able to use appropriate apparatus to make and record a range of measurements accurately.

...... They have no units and only vary from 0-1 ✗

...... **(2 marks)**

2 Simple distillation is used to separate a solvent from a solution.

(a) Suggest reasons that explain why the solution must be heated **gently**.

...... to stop the Solution boiling into the condenser ✓

...... **(2 marks)**

(b) During distillation, the solvent vapour may condense slowly without the use of a condenser. State one hazard caused by carrying out distillation without a condenser.

...... The apparatus gets hot ✓ **(1 mark)**

3 A student investigates the composition of a sample of ballpoint pen ink. He uses propanone for the mobile phase in his paper chromatography. The diagram shows part of the label on a bottle of propanone.

(a) State one hazard of propanone, shown by the label but not described in words.

...... flammable ✓ **(1 mark)**

> ⬥ 🔥 ⬥ !
>
> Irritating to eyes.
> May cause skin dryness.
> Vapour causes dizziness.
> Propanone CH_3COCH_3

(b) Explain two precautions, other than eye protection, to control the risk of harm from propanone.

> You need to be able to use gases, liquids and solids safely and carefully.

......

...... Wear lab coat, Wear Gloves ✓

...... **(2 marks)**

Had a go ☐ Nearly there ☐ Nailed it! ☐

Drinking water

1 Waste water and ground water can be treated to make it safe to drink. Which word correctly describes water that is safe for us to drink?

☒ **A** potable   ☐ **C** edible

☐ **B** fresh   ☐ **D** filtered   **(1 mark)**

2 Chlorine is a toxic gas that dissolves in water.

> Your answer must be more precise than just 'to make it safe to drink'.

(a) State why chlorine is used in water treatment.

...... It kills microbes .... **(1 mark)**

(b) Suggest reasons that explain why drinking water that contains chlorine is considered safe to drink, even though chlorine gas is toxic.

The Concentration of Chlorine is. It kills microbes ...... ✗
but dont harm humans .... **(2 marks)**

3 Name the two stages in water treatment that are carried out before chlorine is added. Give a reason why each stage is carried out.

> You do not have to place the two stages in correct order in this question.

name of stage Sedimentation ....

reason particles sink to the bottom of the tank .....

name of stage filtration ....

reason gets rid of the finer particles .... **(4 marks)**

4 Explain why distilled water, rather than tap water, is used in chemical analysis.

Guided

Unlike tap water, distilled water does not contain microbes dissolved salts .

These would interfere with the results
cost too much and it has high purity costs . **(2 marks)**

5 Drinking water in the UK comes from fresh water including rivers, lakes and reservoirs. In some countries drinking water may come from seawater instead.

(a) Name the separation method used to separate water for drinking from seawater.

...... Simple distillation .... **(1 mark)**

(b) Suggest a reason that explains why producing drinking water from seawater is usually expensive.

...... It has high fuel costs .... **(1 mark)**

6 Aluminium sulfate may be added during water treatment. It forms a precipitate of aluminium hydroxide, which traps small particles suspended in the water. Balance the equation for this reaction.

$Al_2(SO_4)_3(aq) + 6 H_2O(l) \rightarrow 2 Al(OH)_3(s) + 3 H_2SO_4(aq)$ **(1 mark)**

Extended response – Separating mixtures

A cloudy pale yellow mixture contains three substances (**A**, **B** and **C**).

The table shows some information about these substances.

Substance	Melting point (°C)	Boiling point (°C)	Notes
A	115	445	yellow, insoluble in **B** and **C**
B	−95	56	colourless, soluble in **C**
C	0	100	colourless, soluble in **B**

Devise a method to produce pure samples of each individual substance from the mixture.

> You will be more successful in extended response questions if you plan your answer before you start writing.
>
> The command word **devise** means that you are being asked to plan or invent a procedure from existing principles or ideas. You do not have to imagine a complex method that goes beyond your GCSE studies. You should be able to describe an appropriate experimental technique to separate a mixture if you know the properties of the components of the mixture.
>
> You should use the information in the table in your answer, and explain why you have suggested each step.
>
> The separation methods covered at GCSE include simple distillation, fractional distillation, filtration, crystallisation and paper chromatography. You do not need to use them all to answer this question.

....

....

....

....

....

....

....

....

....

....

....

....

.... **(6 marks)**

Acids and alkalis

1   A student adds a few drops of universal indicator solution to some dilute hydrochloric acid.

(a) The universal indicator solution is green before being added to the acid. State what this tells you about the pH of the universal indicator solution.

.... it is neutral .... **(2 marks)**

(b) Give the colour of the mixture formed by the universal indicator solution and dilute hydrochloric acid.

.... **(1 mark)**

(c) Name the ion, produced by the hydrochloric acid, that is responsible for the colour change.

.... **(1 mark)**

2   A student heats some magnesium ribbon in air. It burns with a white flame and a white solid forms. The student then mixes the white solid with water in a test tube.

(a) Write a balanced equation for the reaction between magnesium and oxygen.

> You should be able to recall the formulae of elements and simple compounds.

.... **(2 marks)**

(b) Universal indicator solution turns purple when it is added to the mixture in the test tube. State what this tells you about the mixture.

.... **(1 mark)**

3   Sodium hydroxide dissolves in water to form an alkaline solution.

Write a balanced equation, including state symbols, for the ionisation of sodium hydroxide in solution.

> Two different ions in aqueous solution form.

NaOH(aq) → Na... **(1 mark)**

4   Complete the table below to show the colours of litmus, methyl orange and phenolphthalein.

> You should be able to recall the effect of acids and alkalis on these indicators.

Indicator	Colour at pH 14	Colour at pH 1
litmus	blue	red
methyl orange		
phenolphthalein		

(3 marks)

5   The pH of an acidic solution depends on the concentration of H^+(aq) ions.

(a) State what happens to the pH of an acidic solution as the concentration of these ions is increased.

.... **(1 mark)**

(b) State what happens to the pH of an alkaline solution as the hydroxide ion concentration is increased.

.... **(1 mark)**

Strong and weak acids

1 A student adds water to a sample of an acid. The pH changes from 2 to 4. Which of the following statements about this change is correct?

> Answer **A** cannot be correct because the solution becomes less strongly acidic.

   ☐ **A**   The solution becomes more strongly acidic.

   ☐ **B**   The concentration of hydrogen ions increases 2 times.

   ☐ **C**   The concentration of hydrogen ions increases 100 times.

   ☐ **D**   The concentration of hydrogen ions decreases 100 times.   **(1 mark)**

2 Nitric acid dissociates in aqueous solution: $HNO_3(aq) \rightarrow H^+(aq) + NO_3^-(aq)$

Guided

State why nitric acid is described as a **strong** acid.

Nitric acid is .... dissociated into ions in aqueous solution.   **(1 mark)**

3 Methanoic acid dissociates in aqueous solution: $HCOOH(aq) \rightleftharpoons HCOO^-(aq) + H^+(aq)$

(a) Give the meaning of the symbol $\rightleftharpoons$ in the equation.

.... **(1 mark)**

(b) State why methanoic acid is described as a **weak** acid.

> The reason is **not** to do with the concentration of the methanoic acid.

.... **(1 mark)**

4 Explain, in terms of the amount of dissolved sodium hydroxide, the difference between concentrated sodium hydroxide solution and dilute sodium hydroxide solution.

....

....

.... **(2 marks)**

5 The table gives some information about two different acids.

(a) Explain how the table shows that hydrochloric acid is a stronger acid than ethanoic acid.

Acid	pH of acid
0.2 mol dm^{-3} hydrochloric acid	0.70
0.02 mol dm^{-3} hydrochloric acid	1.70
0.2 mol dm^{-3} ethanoic acid	2.75

....

.... **(2 marks)**

(b) Describe the effect of diluting hydrochloric acid on its pH.

....

.... **(2 marks)**

(c) State why 0.0018 mol dm^{-3} of hydrochloric acid has the same pH as 0.2 mol dm^{-3} of ethanoic acid.

.... **(1 mark)**

Bases and alkalis

1 What forms when an acid reacts with a metal hydroxide?

☐ **A** a salt only

☐ **B** a salt and water only

☐ **C** a salt and hydrogen only

☐ **D** a salt and carbon dioxide only **(1 mark)**

2 Sodium carbonate, Na_2CO_3, reacts with dilute nitric acid:

sodium carbonate + nitric acid → sodium nitrate + carbon dioxide + water

Guided

(a) Write the balanced equation for this reaction.

.... + ... HNO_3 → + ... + ... + ... **(2 marks)**

(b) State two things that you would see when sodium carbonate powder is added to dilute nitric acid.

> You do not need to name any substances in your answer – make sure that you write down two observations.

....

.... **(2 marks)**

(c) Describe the chemical test for carbon dioxide.

> Write down what you would do and what you would expect to observe.

....

.... **(2 marks)**

3 Calcium reacts with dilute hydrochloric acid. Bubbles of gas are given off and a colourless solution forms. Some calcium remains in the bottom of the test tube when the reaction stops.

(a) Name the colourless solution that forms.

.... **(1 mark)**

(b) (i) Name the gas responsible for the bubbles.

.... **(1 mark)**

(ii) Describe the chemical test for the gas named in part (**i**).

....

.... **(2 marks)**

4 Zinc oxide is an example of a base.

(a) Describe what is meant by a **base**.

....

.... **(2 marks)**

(b) State the general name for a **soluble** base.

.... **(1 mark)**

(c) Name the salt formed when zinc oxide reacts with dilute sulfuric acid.

.... **(1 mark)**

Had a go ☐   Nearly there ☐   Nailed it! ☐   **Chemistry Paper 3**

Neutralisation

1 An acid reacts with a base to form a salt and water only. Explain, in terms of reacting ions, what happens when an acid reacts with an alkali.

> Give the names or formulae of the reacting ions from the acid and alkali, and the product that they form.

....

....

.... **(3 marks)**

2 Write balanced equations for the reaction of dilute hydrochloric acid, HCl, with:

(a) calcium oxide powder, CaO

.... **(2 marks)**

(b) calcium hydroxide powder, $Ca(OH)_2$

.... **(2 marks)**

3 Suggest a reason that explains why a pH meter must be calibrated before it is used.

> You should be able to use appropriate apparatus and substances to measure pH in different situations.

.... **(1 mark)**

Guided

4 Limewater is calcium hydroxide solution. A student investigates what happens to the pH when he adds small portions of limewater to 25 cm^3 of dilute hydrochloric acid in a flask. The table shows his results.

Volume of limewater added (cm^3)	pH of the mixture in the flask
0	1.6
5	1.8
10	2.0
15	2.2
20	2.6
24	3.8
25	7.0
26	10.4
30	11.2
35	11.5
40	11.6

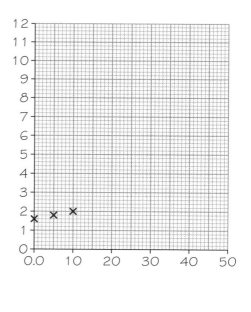

Plot a graph to show these results.   **(3 marks)**

> **Maths skills** If you are asked to plot a graph, you need to mark the points accurately on the grid, then draw a line or curve of best fit. You must also work out a suitable scale and label the axes if this has not already been done for you.

123

Practical skills Salts from insoluble bases

1 Iron(II) oxide is an insoluble base. It reacts with dilute sulfuric acid to form iron(II) sulfate and water:

$$FeO(s) + H_2SO_4(aq) \rightarrow FeSO_4(aq) + H_2O(l)$$

This reaction can be used to prepare a solution that contains only the salt and water.

(a) Explain why an excess of iron(II) oxide is added to the dilute sulfuric acid.

> What substances are in the reaction mixture after adding the excess reactant?

....

.... **(2 marks)**

(b) Suggest a reason that explains why the dilute sulfuric acid may be warmed before adding iron(II) oxide.

.... **(1 mark)**

(c) Name the separation method needed to remove the excess iron(II) oxide.

.... **(1 mark)**

(d) Name the process used to produce iron(II) sulfate crystals from the iron(II) sulfate solution.

.... **(1 mark)**

2 A student wants to prepare pure, dry crystals of copper sulfate. This is her method.

> **Making copper sulfate crystals**
> _____
>
> **A** Put 25 cm³ of dilute sulfuric acid in a beaker.
> **B** Add several spatulas of copper oxide powder together.
> **C** Pour the liquid from the beaker into an evaporating basin.
> **D** Heat the liquid using a blue Bunsen burner flame until all the water has boiled away.

(a) Name a suitable piece of apparatus to measure 25 cm³ of dilute sulfuric acid at step A.

.... **(1 mark)**

(b) Describe **two** improvements the student could make at step B, and give reasons for your answers.

improvement 1 ....

....

improvement 2 ....

.... **(4 marks)**

(c) The method used at step D produces poorly formed, small crystals. Describe how the student should modify the method and apparatus used at step D to produce larger, well-formed crystals safely.

> You should be able to use appropriate heating methods including use of a Bunsen burner and a water bath.

....

....

.... **(2 marks)**

Salts from soluble bases

1 Which of the following is a suitable method to prepare a soluble salt from an acid and an alkali?

.... nitric acid

☐ **A** precipitation

☐ **B** filtration

☐ **C** titration

☐ **D** distillation **(1 mark)** ....

sodium hydroxide + indicator

2 The diagram shows the apparatus used to add known volumes of dilute nitric acid to a measured volume of sodium hydroxide solution.

Complete the diagram to show the names of the pieces of apparatus shown. **(2 marks)**

3 A student prepares sodium chloride solution using the apparatus shown in question **2**.

(a) Name the acid that the student should use in his experiment.

.... **(1 mark)**

(b) Name a piece of apparatus, more accurate than a measuring cylinder, that the student could use to measure 25.0 cm³ of sodium hydroxide solution.

.... **(1 mark)**

(c) The student uses phenolphthalein indicator. Describe the expected colour change at the end-point.

> You need to give the colour at the start and at the end.

.... **(1 mark)**

(d) The student carries out a rough run, then three accurate runs. The table shows his results.

Run number	Rough	1	2	3
End reading (cm³)	26.20	24.90	49.30	24.70
Start reading (cm³)	0.10	0.00	24.90	0.20
Titre (cm³)	26.10			

(i) Suggest a reason that explains why the student carries out a rough run first.

.... **(1 mark)**

Guided

(ii) Complete the table to show the titres for all four runs. **(1 mark)**

(iii) Calculate the mean titre from the accurate runs, ignoring any anomalous (outlier) titre.

.... cm³ **(2 marks)**

(e) Describe how the student should use his mean titre when preparing pure sodium chloride solution.

....

....

.... **(2 marks)**

Making insoluble salts

1   Potassium chloride is a metal chloride that is soluble in water. Which of the following metal chlorides is insoluble in water?

> You need to be able to recall the general rules for the solubility of common types of substances in water.

☐ **A**   sodium chloride

☐ **B**   silver chloride

☐ **C**   copper chloride

☐ **D**   zinc chloride   **(1 mark)**

2   Which of the following pairs contains one substance that is soluble in water and one that is insoluble in water?

☐ **A**   lead chloride and barium sulfate

☐ **B**   calcium nitrate and potassium hydroxide

☐ **C**   aluminium hydroxide and copper carbonate

☐ **D**   ammonium carbonate and calcium sulfate   **(1 mark)**

3   A student wants to produce insoluble calcium hydroxide.

(a) Name two solutions that, when mixed together, will produce a precipitate of calcium hydroxide.

solution 1 ....

solution 2 ....   **(2 marks)**

(b) Name the other product formed when the two solutions named in part (**a**) are mixed together.

....   **(1 mark)**

4   Sodium carbonate, Na_2CO_3, and calcium chloride, $CaCl_2$, are soluble in water. Calcium carbonate, $CaCO_3$, is insoluble in water.

(a) Write a balanced equation for the reaction between sodium carbonate solution and calcium chloride solution. Include state symbols in your answer.

> There are two products of this reaction, including calcium carbonate.

....   **(2 marks)**

(b) Describe how you would use solid sodium carbonate and solid calcium chloride to produce a pure, dry sample of calcium carbonate.

> Remember that you are starting with solid reactants.

....

....

....

....

....   **(4 marks)**

5   Explain why a precipitate forms when dilute sulfuric acid and lead nitrate solution are mixed together.

> Guided

Sulfuric acid contains .... ions which react with .... ions

to form ....   **(3 marks)**

Extended response – Making salts

Sodium chloride solution can be made from dilute hydrochloric acid and sodium hydroxide solution:

$$HCl(aq) + NaOH(aq) \rightarrow NaCl(aq) + H_2O(l)$$

Devise a titration experiment to find the exact volume of hydrochloric acid needed to neutralise 25.0 cm^3 of sodium hydroxide solution. Explain how you would use the result from this experiment to obtain pure, dry, sodium chloride crystals.

> You will be more successful in extended response questions if you plan your answer before you start writing. For example, you could divide your answer here into three sections:
>
> • setting up the apparatus ready for a titration, including where the reagents need to go
>
> • carrying out the titration, including steps needed to obtain an accurate result
>
> • producing sodium chloride crystals from sodium chloride solution.
>
> You should be able to describe how to carry out an acid–alkali titration using a burette, a pipette and a suitable indicator, to prepare a pure, dry salt.

....

....

....

....

....

....

....

....

....

....

....

....

.... (6 marks)

Electrolysis

1   Under what conditions can an ionic compound conduct electricity?

> Molten substances are in the liquid state.

☐ **A** only when it is molten

☐ **B** when it is solid or molten

☐ **C** when it is solid or in solution

☑ **D** when it is molten or in solution

(1 mark)

2   State what is meant by the term **electrolyte**.

Guided

An electrolyte is an ..*ionic*............... compound in the .....*molten*............. state

or ....*dissolved in water*... **(2 marks)**

3   A student places a purple crystal of potassium manganate(VII), $KMnO_4$, on a damp piece of filter paper. She connects each end of the paper to a d.c. electricity supply. A purple streak gradually moves to the left.

Explain why the purple streak moved to the left.

> The potassium ion, K^+, is colourless. What other ion must be present in potassium manganate(VII)?

.......*Ions can move in liquids and in solutions*............... **(2 marks)**

4   In an electrolysis experiment, molten zinc bromide is decomposed.

Predict the product that forms at each electrode.

cathode ..~~Lead~~..*Zinc*..

anode ..*bromine*... **(2 marks)**

5   Sodium is extracted from molten sodium chloride, NaCl, by electrolysis: $Na^+ + e^- \rightarrow Na$

(a) State at which electrode, anode or cathode this reaction happens.

..........~~anode cathode~~...*anode*........................... **(1 mark)**

(b) Explain whether sodium ions are oxidised or reduced in this reaction.

......*The Sodium is being reduced as the positively*......

...*charged ions gain electrons*................................. **(2 marks)**

6   Aluminium is extracted from aluminium oxide, Al_2O_3, by electrolysis.

(a) Aluminium ions, Al^{3+}, form aluminium, Al. Write the balanced half equation for the reaction.

.................$Al^{3+} + 3e^- \rightarrow Al$....................... **(1 mark)**

(b) Oxide ions, O^{2-}, form oxygen gas, O_2. Write the balanced half equation for the reaction.

.................$O^{2-} + 2e^+ \rightarrow O$....................... **(2 marks)**

Electrolysing solutions

1 The ions in copper chloride solution are:

- copper ions, Cu^{2+}
- chloride ions, Cl^-
- hydrogen ions, H^+
- hydroxide ions, OH^-

Copper chloride solution is electrolysed using a d.c. electricity supply.

(a) Which of these ions will be attracted to the cathode during the electrolysis of copper chloride solution?

☐ **A** Cl^- ions only   ☐ **C** H^+ ions only

☐ **B** Cl^- ions and OH^- ions   ☐ **D** H^+ and Cu^{2+} ions   **(1 mark)**

> **Guided**

(b) Explain, with the help of an equation, why the copper chloride solution contains H^+ ions and OH^- ions.

Some water molecules ....

.... **(2 marks)**

(c) Chloride ions, Cl^-, form chlorine gas, Cl_2. Write the balanced half equation for the reaction.

.... **(2 marks)**

2 The electrolysis of concentrated sodium chloride solution, $NaCl(aq)$, produces two useful gases.

(a) Write the formulae of all the ions present in a concentrated sodium chloride solution.

.... **(2 marks)**

(b) Predict the gas that forms at:

(i) the anode .... **(1 mark)**

(ii) the cathode .... **(1 mark)**

(c) Explain why the solution remaining after electrolysis is alkaline.

....

.... **(2 marks)**

3 Oxygen is produced at the anode during the electrolysis of sodium sulfate solution.

Explain how the oxygen is formed from ions in the solution.

> Which negatively charged ions will be present in this solution? Which of these will be discharged at the anode? You could answer in words or using a balanced half equation.

....

.... **(2 marks)**

4 Hydrogen and oxygen are produced during the electrolysis of water. Suggest a reason that explains why electrolysis happens faster when the water is acidified with dilute sulfuric acid.

....

.... **(2 marks)**

Had a go ☐ Nearly there ☐ Nailed it! ☐

 **Practical skills** **Investigating electrolysis**

1 The products formed from copper sulfate solution, $CuSO_4(aq)$, depend on the type of electrode used:

- Graphite electrodes are inert electrodes – they just provide a surface for electrode reactions to happen.

- Copper electrodes are non-inert electrodes – they take part in electrode reactions with copper ions.

(a) Oxygen forms during the electrolysis of copper sulfate solution using graphite electrodes.

Explain at which electrode (anode or cathode) oxygen will be produced.

....

.... **(2 marks)**

(b) The anode loses copper during the electrolysis of copper sulfate solution with copper electrodes.

Write the balanced half equation for the formation of copper ions, Cu^{2+}, from copper.

.... **(2 marks)**

2 The electrolysis of copper sulfate solution, using copper electrodes, is used to purify copper. During electrolysis, the copper anode loses mass. The copper cathode gains mass because copper is deposited.

(a) Write the balanced half equation for the formation of copper from copper ions, Cu^{2+}.

.... **(2 marks)**

(b) A student investigates the gain in mass by a copper cathode. She runs each experiment for the same time, but changes the current. She measures the mass of the cathode before and after electrolysis. The graph shows her results.

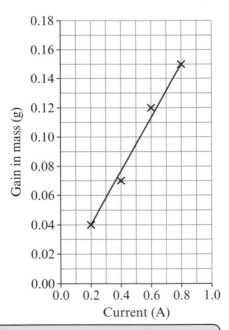

(i) Identify the variable controlled by the student in her experiment.

....

....**(1 mark)**

(ii) Identify the dependent variable in the student's experiment.

....

....**(1 mark)**

(iii) Calculate the gradient of the line of best fit. Give your answer to two significant figures.

A linear relationship such as this can be represented by: $y = mx + c$ (m is the gradient and c is the intercept on the vertical axis (y-axis). The gradient equals the change on the y-axis, divided by the change on the x-axis.

.... g/A **(3 marks)**

Extended response – Electrolysis

A student carries out two experiments using copper chloride, $CuCl_2$.

In experiment 1, the student places two graphite electrodes into copper chloride powder in a beaker. She then connects the electrodes to a d.c. electricity supply and records any changes.

For experiment 2, the student disconnects the d.c. supply, then adds some water to dissolve the copper chloride. She reconnects the electrodes to the d.c. supply and records any changes.

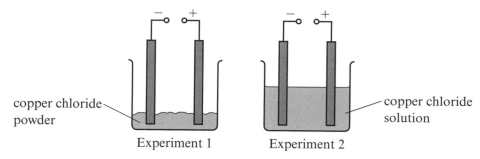

copper chloride powder   copper chloride solution

Experiment 1   Experiment 2

The table shows the student's results.

Experiment	Observations at the cathode (−)	Observations at the anode (+)
1	no visible change	no visible change
2	brown solid forms on the electrode	bubbles of a yellow-green gas released

Explain the differences between the results shown in the table for experiments 1 and 2.

> You will be more successful in extended response questions if you plan your answer before you start writing.
>
> Explain why copper chloride powder does not conduct electricity, and then explain why copper chloride solution does conduct electricity. Name the substances responsible for the student's observations. You should explain, in terms of ions, why each substance forms. You could include half equations for this.

....

....

....

....

....

....

....

....

....

....

....

.... **(6 marks)**

Continue your answer on your own paper. You should aim to write about half a side of A4.

The reactivity series

1 Four metals (**W**, **X**, **Y** and **Z**) are added to cold water and to dilute hydrochloric acid. The table shows what happens.

Metal	Observations in water	Observations in dilute hydrochloric acid
W	slow bubbling	very fast bubbling
X	no visible change	no visible change
Y	fast bubbling	very fast bubbling
Z	no visible change	slow bubbling

(a) Which of the following shows the order of reactivity, from most reactive to least reactive metal?

☐ **A** W, Y, X, Z   ☐ **C** Y, W, X, Z

☐ **B** X, Z, W, Y   ☐ **D** Y, W, Z, X   **(1 mark)**

(b) The concentration of hydrochloric acid is kept the same each time. Give two other variables that should be kept the same in each experiment so that the reactivity of the metals can be compared.

....

....   **(2 marks)**

2 Magnesium reacts slowly with cold water to produce magnesium hydroxide, $Mg(OH)_2$.

> This flammable gas is produced when any metal reacts with water or with dilute acids.

(a) Name the gas produced in the reaction.

....   **(1 mark)**

(b) Write the balanced equation for the reaction between magnesium and water.

....   **(2 marks)**

(c) Name the compound formed when magnesium reacts with steam, rather than with cold water.

....   **(1 mark)**

3 Aluminium is protected from contact with water by a natural layer of aluminium oxide, Al_2O_3. This means that aluminium does not react with water, even though it is a reactive metal. However, aluminium does react with dilute acids, such as dilute sulfuric acid.

⟩ **Guided** ⟩

(a) Write the balanced equation for the reaction between aluminium oxide and dilute sulfuric acid.

Al_2O_3 + ...H_2SO_4 → $Al_2(SO_4)_3$ + ...H_2O **(1 mark)**

(b) Aluminium reacts with dilute sulfuric acid to form aluminium sulfate solution and a flammable gas.

Write the balanced equation for this reaction. Include state symbols in your answer.

....   **(3 marks)**

(c) There is no immediate visible change when aluminium is added to dilute sulfuric acid. Bubbling then starts and gets increasingly faster. Suggest reasons that explain these observations.

....

....

....   **(2 marks)**

Metal displacement reactions

1 Copper can displace silver from silver nitrate solution. Copper nitrate solution also forms in the reaction.

    (a) Give a reason that explains why copper can displace silver from silver salts in solution.

> Which metal, copper or silver, is the more reactive of the two metals?

.... **(1 mark)**

Guided

    (b) Write the balanced equation for the reaction between copper and silver nitrate solution. Include state symbols in your answer.

    ....(.......) + ...$AgNO_3$(.......) → ....(......) + $Cu(NO_3)_2$(.......)   **(3 marks)**

2 A student investigates the reactivities of four metals, copper, magnesium, zinc and X. She adds pieces of magnesium ribbon to solutions of the nitrates of each metal. She then removes and examines each piece of magnesium ribbon after a few minutes. The table shows her results.

Solution	Observations
copper nitrate	brown coating on the magnesium ribbon
magnesium nitrate	no visible change
zinc nitrate	no visible change
X nitrate	grey coating on the magnesium ribbon

    (a) Name the substance found in the brown coating on the magnesium ribbon.

.... **(1 mark)**

    (b) Give a reason that explains why there is no visible change when magnesium nitrate solution is used.

.... **(1 mark)**

    (c) The student repeats the experiment but she uses pieces of metal X instead of magnesium ribbon. The table shows her results.

Solution	Observations
copper nitrate	brown coating on the piece of metal X
magnesium nitrate	no visible change
zinc nitrate	grey coating on the piece of metal X
X nitrate	no visible change

Use the results shown in both tables to place the four metals in order of **decreasing** reactivity.

most reactive ....

....

....

least reactive .... **(2 marks)**

3 The thermite reaction makes molten iron for welding railway lines:

$$2Al(s) + Fe_2O_3(s) \rightarrow Al_2O_3(s) + 2Fe(l)$$

Explain what this reaction shows about the relative reactivity of aluminium and iron.

....

.... **(2 marks)**

Explaining metal reactivity

1 Give the meaning of the term **cation**.

> **Guided**

A cation is a .... charged ion.   **(1 mark)**

2 Calcium is a reactive metal. It reacts vigorously with dilute hydrochloric acid to form calcium chloride solution and hydrogen gas: $Ca(s) + 2HCl(aq) \rightarrow CaCl_2(aq) + H_2(g)$

(a) The formula for a chloride ion is Cl^-. Deduce the formula for a calcium ion.

.... **(1 mark)**

(b) Describe what happens when a calcium atom becomes a calcium ion.

> The outer shell is involved when atoms of an element take part in reactions.

....

.... **(2 marks)**

(c) The table shows a reactivity series for the metals. Hydrogen is a non-metal. It is included for comparison.

> The more easily a metal's atoms form cations, the more reactive the metal is.

Identify the metal that:

(i) forms cations most easily

.... **(1 mark)**

(ii) forms cations least easily.

.... **(1 mark)**

potassium	most reactive
sodium	
calcium	
magnesium	
aluminium	
zinc	
iron	
(hydrogen)	
copper	
silver	
gold	least reactive

(d) Identify a metal that will **not** react with dilute acids.

.... **(1 mark)**

3 Metal displacement reactions are **redox** reactions.

(a) Zinc displaces copper from copper sulfate solution: $Zn(s) + CuSO_4(aq) \rightarrow ZnSO_4(aq) + Cu(s)$

Explain this reaction in terms of the tendency to form cations.

....

.... **(2 marks)**

> **Guided**

(b) Write balanced half equations for:

(i) the oxidation of magnesium to form magnesium ions in aqueous solution

$Mg(s) \rightarrow$ .... **(2 marks)**

(ii) the reduction of hydrogen ions in an acidic solution to form hydrogen gas.

........$H^+(aq) +$ .... $\rightarrow$ .... **(2 marks)**

Metal ores

1   Tungsten metal is extracted from tungsten oxide. The tungsten oxide is heated in a stream of hydrogen gas:

$$WO_3 + 3H_2 \rightarrow W + 3H_2O$$

(a) What happens in this reaction?

☐ **A**   Tungsten is oxidised.   ☐ **C**   Hydrogen is reduced.

☐ **B**   Tungsten oxide is reduced.   ☐ **D**   Water is oxidised.   **(1 mark)**

(b) Suggest a reason that explains why this process is hazardous.

> Look at the reactants and products – are any of them hazardous?

.... **(1 mark)**

2   Metals are extracted from their ores. Give the meaning of the term **ore**.

> Guided

a rock or mineral that contains ....

.... **(2 marks)**

3   Explain why some metals are found in the Earth's crust as uncombined elements.

....

.... **(2 marks)**

4   Cassiterite is a tin ore that contains tin oxide, SnO_2. Tin is extracted from tin oxide by heating with powdered carbon. Carbon monoxide, CO, also forms in the reaction.

(a) Write a balanced equation for the reaction.

.... **(2 marks)**

(b) Explain, in terms of gain or loss of oxygen, whether tin oxide is oxidised or reduced.

....

.... **(2 marks)**

5   Corrosion occurs when a metal oxidises, and this process continues. For example, sodium is shiny when it is freshly cut, but a dull layer of sodium oxide, Na_2O, forms quickly when sodium is exposed to air.

(a) Write a balanced equation for the reaction between sodium and oxygen gas, forming sodium oxide.

.... **(2 marks)**

(b) Suggest a reason that explains why copper oxidises slowly unless it is heated strongly.

.... **(1 mark)**

6   One of the stages in the extraction of titanium metal involves heating titanium chloride with sodium:

$$TiCl_4 + 4Na \rightarrow Ti + 4NaCl$$

(a) The formula for a chloride ion is Cl^-. Deduce the formula for a titanium ion.

.... **(1 mark)**

(b) Explain, in terms of gain or loss of electrons, whether titanium ions are oxidised or reduced.

....

.... **(2 marks)**   **135**

Iron and aluminium

1 The table shows a reactivity series for the metals.

Carbon is a non-metal. It is included for comparison.

Name a metal in the table that:

(a) is likely to be extracted from its ore using electrolysis.

.... **(1 mark)**

sodium	most reactive
calcium	
magnesium	
(carbon)	↓
zinc	
copper	least reactive

(b) could be extracted from its ore by heating with carbon.

.... **(1 mark)**

2 Haematite is an iron ore. It contains iron(III) oxide, Fe_2O_3.

Describe how iron is extracted from this ore. In your answer, include a balanced equation for the reaction.

Iron is placed between zinc and copper in the reactivity series.

....

.... **(3 marks)**

3 Bauxite is an aluminium ore. It contains aluminium oxide, Al_2O_3.

Aluminium is extracted from purified aluminium oxide by electrolysis.

Guided

(a) Explain why aluminium oxide must be molten or dissolved for electrolysis to occur.

The ions in the electrolyte must be ....

.... **(2 marks)**

(b) Aluminium oxide is dissolved in molten cryolite, rather than being heated to melt it.

Suggest a reason that explains why this is done.

Think about the amounts of energy involved in each case.

.... **(1 mark)**

(c) Write half equations for the reactions that happen at each electrode during the electrolysis of aluminium oxide.

at the anode ....

at the cathode .... **(4 marks)**

4 Aluminium is more abundant than iron in the Earth's crust. On average, the crust contains 8.2% aluminium but only 6.3% iron, although aluminium costs about six times more than iron.

What is likely to be the main reason why extracting aluminium is more expensive than extracting iron?

Suggest reasons that explain the difference in the cost of the two metals.

....

.... **(2 marks)**

Biological metal extraction

1 Most **high-grade** copper ores have been used up. This means that copper must be extracted from **low-grade** ores. These have only low concentrations of copper compounds.

Phytoextraction is one way to extract copper from low-grade ores.

Guided

(a) The table shows the main steps involved in phytoextraction. Give the correct order by writing the numbers 1 to 5 in the correct boxes, where step 1 is the first step.

Step number	Process
	Copper ions become concentrated in compounds in plants.
	Copper is extracted from ash with a high concentration of copper compounds.
	Plants absorb copper ions through their roots.
1	Sow plants on ground containing low-grade copper ore.
	Plants are harvested and burned.

(3 marks)

(b) Energy is transferred to the environment, mainly by heating, when plants are burned. Suggest reasons that explain why this may be useful to a company extracting copper by phytoextraction.

....

.... **(2 marks)**

(c) Describe a disadvantage of extracting copper by phytoextraction.

....

.... **(1 mark)**

2 Bioleaching is one way to extract copper from low-grade ores. Reactions involving bacteria cause the slow conversion of copper sulfide to a mixture of copper sulfate solution and sulfuric acid.

(a) Suggest a reason that explains why the formation of sulfuric acid may be harmful to the environment.

.... **(1 mark)**

(b) The copper sulfate solution is very dilute. Scrap iron is used to produce copper from this solution.

(i) Write a balanced equation for the reaction between iron and copper sulfate solution, forming iron(II) sulfate solution and copper. Include state symbols in your answer.

> Copper ions have the formula Cu^{2+}, iron(II) ions have the formula Fe^{2+} and sulfate ions are SO_4^{2-}.

.... **(3 marks)**

(ii) State why iron is able to displace copper in this reaction.

.... **(1 mark)**

(iii) Give an advantage of using scrap iron to produce copper.

> Which metal is more valuable, copper or iron?

....

.... **(1 mark)**

Recycling metals

1 Some metals are found uncombined in the Earth's crust, but most metals are found as compounds in rocks.

(a) What name is given to any rocks from which metals can be extracted?

> Answer D cannot be correct because quarries are places where these rocks are removed from the ground.

☐ **A** oxides

☐ **B** ores

☐ **C** limestone

☐ **D** quarries   **(1 mark)**

(b) Describe two ways in which large-scale removal of rocks from the ground damages the environment.

....

.... **(2 marks)**

2 Around 90% of the lead produced each year is used in traditional 'lead acid' batteries for cars and other vehicles. About 70% of the lead used each year is recycled lead.

Guided

(a) Describe an advantage of recycling lead from lead acid batteries, rather than recycling lead from general scrap metal waste.

> Think about how different metals are obtained from scrap metal waste.

Most lead for recycling is found in ....

so lead does not need to be .... **(2 marks)**

(b) Describe two advantages of recycling metals, rather than extracting them from their compounds.

....

.... **(2 marks)**

3 Food cans are made from either steel coated with tin or aluminium. The table shows some information about these three metals.

Metal	Abundance in the Earth's crust (%)	Cost of 1 tonne of metal (£)	Energy saved by recycling (%)
steel	6.3	500	70
tin	0.000 22	16 500	75
aluminium	8.2	1500	94

(a) Some people believe that it may be more important to recycle tin, rather than steel or aluminium. Justify this statement using information from the table.

> **Maths skills** Use evidence from the table to support this statement.

....

.... **(2 marks)**

(b) The mass of metal used each year should also be taken into account when assessing the advantages of recycling a particular metal. Suggest a reason that explains why this is so.

....

.... **(1 mark)**

Life-cycle assessments

1 A life-cycle assessment for a manufactured product involves considering its effect on the environment at all stages.

The table shows the main steps involved in a life-cycle assessment. Give the correct order by writing the numbers 1 to 4 in the correct boxes, where step 1 is the first.

(1 mark)

Step number	Process
	manufacturing the product
	obtaining raw materials
4	disposing of the product
	using the product

2 Manufacturers can make glass bottles with thinner walls than in the past. A bottle for a fizzy drink had a mass of 240 g in 1996 but a mass of 190 g in 2016.

(a) 16.5 MJ/kg of glass is used in their manufacture.

Calculate the energy, in MJ, needed to make one bottle in 1996.

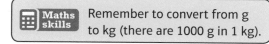

Maths skills — Remember to convert from g to kg (there are 1000 g in 1 kg).

.... MJ **(2 marks)**

(b) Carbon dioxide, CO_2, is a greenhouse gas. The manufacture of glass bottles causes the emission of 1.2 kg CO_2/kg glass. Calculate the difference, between 1996 and 2016, in the mass of carbon dioxide emitted when one bottle is made.

.... **(2 marks)**

3 Window frames may be made from either PVC (a polymer) or wood. The table shows some information from a life-cycle assessment of a window frame.

Process	Energy used (MJ)	
	PVC frame	Wooden frame
producing the material	12.0	4.0
making the frame	3.0	3.6
transport and installation	4.2	4.8
maintenance	0.3	1.5
disposal in landfill	0.7	0.8

(a) Identify the stage in the life cycle of **each** window frame that is responsible for the most energy use.

....

.... **(1 mark)**

(b) Explain which type of frame is likely to have the lower environmental impact **when in use**.

....

.... **(2 marks)**

(c) The disposal of a PVC frame uses less energy than the disposal of a wooden frame. Suggest a reason that explains why this should not be the only measure of the environmental impact of disposal.

What happens to the frames after they have been buried in a landfill site?

.... **(1 mark)**

Extended response – Reactivity of metals

Magnesium forms cations more readily than copper. A spatula of magnesium powder mixed with a spatula of copper oxide powder is heated strongly on a steel lid. Magnesium oxide and copper are produced:

magnesium + copper oxide → magnesium oxide + copper
$$Mg(s) + CuO(s) \rightarrow MgO(s) + Cu(s)$$

steel lid —— —— reaction mixture

tripod ——

HEAT

heat-resistant mat ——

The reaction shows that magnesium is more reactive than copper. It is a very vigorous reaction. Energy is transferred to the surroundings by light and heating, and hot powder escapes into the air.

Devise an experiment, based on this method, to investigate the relative reactivity of copper, iron and zinc. In your answer, describe the results you expect, and explain how you would use them to deduce the order of reactivity. Explain how you would control the risks of harm in the investigation.

> You will be more successful in extended response questions if you plan your answer before you start writing.
>
> Think about how many combinations of a metal powder and a metal oxide powder you will need to test.
>
> One way to show these combinations, and the expected results, is to make a completed results table.
>
> You should be able to evaluate the risks in a practical procedure, and suggest suitable precautions for a range of practicals (not just those mentioned in the specification).

....

....

....

....

....

....

....

....

....

....

....

....

.... **(6 marks)**

The Haber process

1   In the Haber process, nitrogen and hydrogen react together to form ammonia:

$$N_2(g) + 3H_2(g) \rightleftharpoons 2NH_3(g)$$

(a)   What is the raw material for the nitrogen?

☐ **A**   air   ☐ **C**   seawater

☐ **B**   natural gas   ☐ **D**   hydrochloric acid   **(1 mark)**

(b)   Give the meaning of the symbol $\rightleftharpoons$ in the balanced equation.

.... **(1 mark)**

2   The conditions used in the Haber process are carefully controlled to achieve an acceptable yield of ammonia in an acceptable time.

(a)   State the temperature and pressure used in the Haber process.

temperature ....°C

pressure ....atmospheres   **(2 marks)**

(b)   Explain why iron is used in the Haber process.

> The reactor is made from steel, and iron is not a reactant or a product.

....

.... **(2 marks)**

3   Dilute ethanoic acid reacts with ethanol. Ethyl ethanoate and water form in the reaction:

> You may see unfamiliar chemistry in an unfamiliar context, like this one.

$$CH_3COOH(aq) + CH_3CH_2OH(l) \rightleftharpoons CH_3COOCH_2CH_3(aq) + H_2O(l)$$

All four substances are clear, colourless liquids. They mix completely with each other.

(a)   State what visible changes, if any, you would observe during the reaction.

> Do not be put off by the complex appearance of the equation. It is just an example of: $A + B \rightleftharpoons C + D$

.... **(1 mark)**

⟩ **Guided** ⟩   (b)   The reaction reaches a dynamic equilibrium after a few days.

(i)   Describe what is happening to the forward and backward reactions at equilibrium.

The rates of the forward and backward reactions are ....

and they .... **(2 marks)**

(ii)   State what happens to the concentrations of the reacting substances at equilibrium.

> The basic choices you have are: increase, decrease, do not change.

.... **(1 mark)**

More about equilibria

1 The manufacture of sulfuric acid involves a reversible reaction between sulfur dioxide and oxygen:

$$2SO_2(g) + O_2(g) \rightleftharpoons 2SO_3(g) \qquad \text{(forward reaction is exothermic)}$$

(a) Explain the effect of increasing the pressure on the position of equilibrium.

> State what will happen (moves to the left, moves to the right or stays the same), and give a reason why.

....

.... **(2 marks)**

(b) Explain the effect of increasing the temperature on the position of equilibrium.

....

.... **(2 marks)**

(c) State the effect on the position of equilibrium, if the sulfur trioxide is removed.

.... **(1 mark)**

(d) Vanadium(V) oxide, V_2O_5, is used as a catalyst. State the effect of using a catalyst on the position of equilibrium.

.... **(1 mark)**

2 Ammonia is made from nitrogen and hydrogen using the Haber process:

$$N_2(g) + 3H_2(g) \rightleftharpoons 2NH_3(g)$$
(forward reaction is exothermic)

The graph shows how the yield of ammonia depends on the temperature and pressure used.

(a) Identify the temperature and pressure, shown by the graph, needed to obtain the maximum yield.

.... **(1 mark)**

Guided

(b) The conditions chosen are 450 °C and 200 atmospheres pressure (atm).

  (i) Identify the percentage yield of ammonia obtained under these conditions, as shown by the graph.

.... **(1 mark)**

  (ii) Explain why these conditions are described as a **compromise** temperature and pressure.

> Mention the position of equilibrium, the rate of reaction and relevant factors such as cost.

Lower temperatures give a ....

but ....

Higher pressures give a ....

but .... **(4 marks)**

The alkali metals

1 In general, the alkali metals are:

☑ **A** hard with relatively low melting points   ☐ **C** soft with relatively low melting points

☐ **B** soft with relatively high melting points   ☐ **D** hard with relatively high melting points

(1 mark)

2 Give a reason that explains why, in terms of electronic configurations, the alkali metals occupy group 1.

> The electronic configurations of the atoms of these elements differ, but they do have something in common.

..........The alkali's ....Occupy group 1 because they are the most reactive.... **(1 mark)**

3 The alkali metals react with water to produce a metal hydroxide and hydrogen. For example:

sodium + water → sodium hydroxide + hydrogen

> **Guided**

(a) Write the balanced equation for this reaction. Include state symbols in your answer.

..2..Na(.S..) + ...2H$_2$O.(l)... → ..2..NaOH(aq) +..2.. H$_2$..(g).... **(3 marks)**

(b) Explain why, during this reaction, the sodium forms a molten ball.

.........the .....Sodium .....reacts .....With ....water..........

.... **(2 marks)**

(c) Describe the chemical test for hydrogen.

.......The ....Squeaky .....Pop .....test............

> Make sure that you can recall and apply relevant knowledge and understanding from other topics.

.... **(2 marks)**

4 Explain why lithium, sodium and potassium are stored in oil.

........The ....are .....less .....dense ....than ....Wob..

.... **(2 marks)**

5 Describe what you would see when a small piece of potassium is added to water.

> You should not name any products in your answer – just write down what you would **see**.

..It .....gives ....of .....Sparks .....and ....produce..........

...fire..

.... **(3 marks)**

6 Reactivity increases going down group 1, from lithium to potassium. Explain this pattern in reactivity in terms of the electronic configurations of the atoms.

> **Guided**

Going down the group, the size of the atoms...clecrase.......and ....they.....

....Get ....more .....reactive.............................

.... **(3 marks)**

The halogens

1   Which of the following is a chemical test for chlorine gas?

> Answer D cannot be correct because chlorine must dissolve in water for the chemical test to work.

☐ **A** Damp red litmus paper turns blue, then white.

☐ **B** Damp blue litmus paper turns red, then white.

☐ **C** Damp starch iodide paper turns red, then white.

☒ **D** Dry starch iodide paper turns blue–black.   **(1 mark)**

2   Give a reason that explains why, in terms of electronic configurations, the halogens occupy group 7.

....   **(1 mark)**

3   Complete the table to show the colours and physical states of the halogens at room temperature.

> Guided

Halogen	Colour	Physical state
chlorine	Yellow-green	gas
bromine	red-brown	liquid
iodine	dark grey	solid (forms a purple vapour)

(6 marks)

4   The table shows the densities of two halogens, in order going down group 7.

Halogen	Density at room temperature and pressure (kg/m^3)
bromine	3103
iodine	4933

Predict the density of astatine, the element placed immediately below iodine, and justify your answer.

....

....   **(2 marks)**

5   Fluorine at the top of group 7 exists as simple molecules. Each molecule contains two fluorine atoms.

(a) Name the type of bond that exists between the atoms in a fluorine molecule.

....   **(1 mark)**

(b) Explain why fluorine has a low boiling point.

> In your answer, make sure that you identify the bonds or forces overcome during boiling.

....

....   **(2 marks)**

(c) Describe and explain the trend in melting point going down group 7.

....

....   **(2 marks)**

Reactions of halogens

1 In the cold and dark, hydrogen reacts explosively with fluorine to produce hydrogen fluoride, HF.

(a) Write a balanced equation for this reaction.

> Remember that the gaseous elements (apart from those in group 0) exist as diatomic molecules, X_2.

.... **(2 marks)**

(b) What happens when hydrogen fluoride is added to water?

☐ **A** It reacts vigorously, releasing oxygen.

☐ **B** It dissolves to form an alkaline solution.

☐ **C** It dissolves to form an acidic solution.

☐ **D** It dissolves to form a neutral salt solution.   **(1 mark)**

(c) If exposed to sunlight, a mixture of hydrogen and chlorine reacts explosively. Suggest reasons that explain why a mixture of hydrogen and bromine reacts only if a flame is put in it.

> Look again at the reaction conditions needed for hydrogen with fluorine or chlorine.

....

....

.... **(2 marks)**

2 A teacher heats a small piece of sodium in a steel deflagrating spoon until the sodium ignites. She then puts the spoon in a gas jar of chlorine. The sodium burns in the chlorine to produce sodium chloride.

Guided

(a) Write the balanced equation for this reaction. Include state symbols in your answer.

......Na(...) + Cl_2(......) → ......NaCl(......)   **(3 marks)**

(b) The teacher then passes bromine vapour over hot iron wool. Red–brown iron(III) bromide, $FeBr_3$, is produced. Write the balanced equation for this reaction.

.... **(2 marks)**

(c) A student places some iron wool in a boiling tube with a few crystals of iodine. He heats the iodine gently to produce iodine vapour, and then heats the iron wool strongly. The iron and iodine react slowly to produce grey iron(II) iodide, FeI_2.

> Make sure that you can recall the formulae of elements, simple compounds and ions.

(i) Write the formula of the iron(II) ion and the formula of the iodide ion.

.... **(2 marks)**

(ii) Write the balanced equation for this reaction.

.... **(2 marks)**

3 Reactivity decreases going down group 7, from fluorine to iodine. Explain, in terms of the electronic configurations of their atom, why fluorine is more reactive than chlorine.

Guided

Fluorine atoms are .... than chlorine atoms ....

....

.... **(3 marks)**

Halogen displacement reactions

1 A student adds a few drops of aqueous bromine solution to a potassium iodide solution. Iodine and potassium bromide solution form. What type of reaction is this?

> Answer **C** cannot be correct because distillation is a physical separation method, not a chemical reaction.

☐ **A** neutralisation

☐ **B** precipitation

☐ **C** distillation

☐ **D** redox   **(1 mark)**

2 A displacement reaction may happen when a halogen is added to a solution containing halide ions. The table shows results from an investigation with three halogens. A tick (✓) shows that displacement happened.

Halogen added	Halide ion in solution		
	Chloride	**Bromide**	**Iodide**
chlorine	not done	✓	✓
bromine	✗	not done	✓
iodine	✗	✗	not done

> **Guided**

(a) Use the results shown in the table to deduce the order of reactivity of these halogens.

The order of reactivity, starting with the most reactive, is...

.... because chlorine displaces ....

.... but bromine displaces only..............................

Iodine .... **(3 marks)**

(b) Suggest a reason that explains why three possible experiments were not done in the investigation.

.... **(1 mark)**

(c) Predict whether iodine will be able to displace astatine from astatide ions. Explain your answer.

....

.... **(2 marks)**

3 Fluorine displaces iodine from potassium iodide solution soaked into filter paper:

> Potassium ions, K⁺, are **spectator ions**. You can leave them out of the ionic equation.

$$F_2(g) + 2KI(aq) \rightarrow 2KF(aq) + I_2(aq)$$

> **Guided**

(a) Write an ionic equation for this reaction.

$F_2(g) +$ .... **(2 marks)**

(b) Explain, in terms of the gain or loss of electrons, which substance is:

(i) oxidised

.... **(2 marks)**

(ii) reduced

.... **(2 marks)**

The noble gases

1 Which of these properties explains why argon is used as a shield gas during welding?

   ☐ **A** Argon is inert.

   ☐ **B** Argon is flammable.

   ☐ **C** Argon has a low density.

   ☐ **D** Argon is a good conductor of electricity.   **(1 mark)**

2 Explain why helium is used as a lifting gas for party balloons and airships.

   > There are two relevant properties. For each one, explain why it is important for this use of helium.

   ....

   .... **(2 marks)**

3 State, in terms of electronic configurations, why the noble gases occupy group 0 of the period table.

   .... **(1 mark)**

4 The table shows some information about the noble gases.

Element	Melting point (°C)
helium	−272
neon	−248
argon	−189
krypton	−157
xenon	−111
radon	−71

   > Temperatures with less negative numbers are higher temperatures, so −10 °C is warmer than −20 °C.

   (a) Name the noble gas that has the lowest melting point.

   .... **(1 mark)**

   (b) Oganesson, Og, was discovered early this century. It is placed in group 0 of the periodic table, immediately below radon. Predict the melting point of oganesson, and explain your answer.

   ....

   .... **(2 marks)**

5 The electronic configuration of He (atomic number 2) is 2.

   (a) State the electronic configuration of:

       (i) neon (atomic number 10):

   .... **(1 mark)**

       (ii) argon (atomic number 18):

   .... **(1 mark)**

   (b) Explain, in terms of their electronic configurations, why the noble gases are unreactive.

   ....

   .... **(2 marks)**

147

Extended response – Groups

The diagram shows the first five elements in groups 1 and 7 of the periodic table.

Group 1	Group 7
lithium	fluorine
sodium	chlorine
potassium	bromine
rubidium	iodine
caesium	astatine

In 2012, the reaction between caesium and fluorine was filmed for the first time. As predicted, the elements reacted together very violently, producing caesium fluoride:

> Caesium fluoride is an ionic compound, formed when caesium and fluorine form oppositely charged ions.

$$2Cs(s) + F_2(g) \rightarrow 2CsF(s)$$

Explain, in terms of electrons, how caesium and fluorine react together to form caesium fluoride. Explain whether each element is reduced or oxidised, and state why this reaction is very violent.

> The reaction is a redox reaction, in which one element loses electrons and the other element gains electrons.

> You will be more successful in extended response questions if you plan your answer before you start writing.
>
> What are the trends in reactivity in groups 1 and 7?
>
> This question also covers content from Topic 1 (key concepts in chemistry). Remember that this topic is common to Paper 1 and Paper 2, and not just covered in Paper 1.

....

....

....

....

....

....

....

....

....

....

....

....

.... **(6 marks)**

Rates of reaction

1 Magnesium ribbon reacts with dilute hydrochloric acid. Magnesium chloride solution and hydrogen gas are formed. Which of the following changes would cause a decrease in the rate of this reaction?

☐ **A** Use a larger volume of hydrochloric acid.

☐ **B** Dilute the hydrochloric acid with water.

☐ **C** Use warmer dilute hydrochloric acid.

☐ **D** Use magnesium powder instead of ribbon. **(1 mark)**

2 Explain what must happen to reactant particles for a reaction to occur.

....

.... **(2 marks)**

3 (a) Describe the meaning of the term **catalyst**.

a substance that speeds up a reaction without altering the ....

and is unchanged ....

.... **(3 marks)**

(b) Explain, in terms of energy, how a catalyst increases the rate of a reaction.

....

.... **(2 marks)**

(c) (i) State the name given to a biological catalyst.

.... **(1 mark)**

(ii) Give one example of a commercial use of a biological catalyst.

.... **(1 mark)**

4 Lumps of zinc react with dilute sulfuric acid to form zinc sulfate and hydrogen. Explain, using ideas about reactant particles, why the reaction is faster when the same mass of zinc powder is used.

Think about the surface area to volume ratio of the lumps and powder.

....

....

.... **(2 marks)**

5 Copper oxide powder reacts with dilute hydrochloric acid to form copper chloride and water.

The reaction happens faster if the dilute hydrochloric acid is warmed up before adding the powder.

You need to think about the energy of collisions as well as their frequency.

Explain, using ideas about reactant particles, why this happens.

....

....

.... **(3 marks)**

Investigating rates

1   Sodium thiosulfate solution and dilute hydrochloric acid are clear, colourless solutions. They react together to form sodium chloride solution, water, sulfur dioxide and sulfur:

$$Na_2S_2O_3(aq) + 2HCl(aq) \rightarrow 2NaCl(aq) + H_2O(l) + SO_2(g) + S(s)$$

(a) Suggest reasons that explain why the production of sodium chloride solution or water **cannot** easily be used to determine the rate of reaction.

....

.... **(2 marks)**

(b) Sulfur dioxide is a gas that is highly soluble in water.

  (i)   Give one way in which the volume of a gas can be measured accurately.

.... **(1 mark)**

  (ii) Explain why measuring sulfur dioxide is **not** a reliable way to determine the rate of this reaction.

....

.... **(2 marks)**

2   A student investigates how changes in the concentration affect the rate of the reaction between sodium thiosulfate solution and dilute hydrochloric acid. She uses the method shown in the diagram.

The student varies the concentration of sodium thiosulfate solution by diluting it with water. She uses $5\ cm^3$ of $2.0\ mol\ dm^{-3}$ hydrochloric acid each time. The table shows her results.

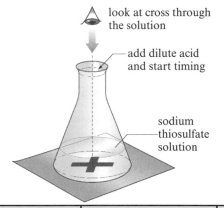

look at cross through the solution

add dilute acid and start timing

sodium thiosulfate solution

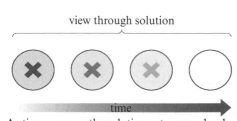

view through solution

time

As time goes on, the solution gets more cloudy. The cross 'disappears'.

Volume of $0.20\ mol\ dm^{-3}$ $Na_2S_2O_3(aq)$ (cm^3)	Volume of water added (cm^3)	Concentration of $Na_2S_2O_3(aq)$ added $(mol\ dm^{-3})$	Time taken for cross on paper to disappear (s)	Relative rate of reaction, $1000 \div$ time (/s)
10	40	0.04	125	
30	20	0.12	42	24
50	0	0.20	25	

(a) Describe how the student controls the volume of sodium thiosulfate used in each experiment.

Study the volumes of sodium thiosulfate solution and water given in the table.

.... **(1 mark)**

Guided   (b) Complete the table to show the relative rate of each reaction.   **(3 marks)**

(c) Describe the relationship between the relative rate of reaction and the concentration of $Na_2S_2O_3(aq)$.

....

.... **(2 marks)**

Skills – Rates of reaction

1   A student investigates the rate of reaction between calcium carbonate and dilute hydrochloric acid. He adds some small lumps of calcium carbonate to an excess of acid in a flask, and measures the change in mass. The table shows the student's results.

Time (s)	Change in mass (g)
0	0.00
20	0.48
40	0.76
60	0.88
80	0.94
100	0.96
120	0.96

(a)   Plot a graph of change in mass against time using the grid. **(3 marks)**

> Use × or + for each point, and draw a single line of best fit. The line does not have to be a straight line.

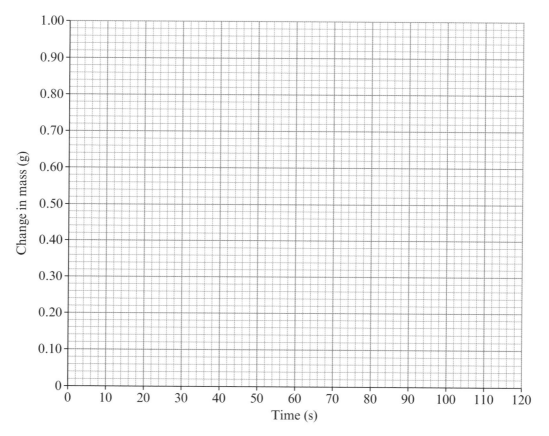

(b)   Describe how the student can tell from his results that the reaction has finished.

....

.... **(1 mark)**

(c)   The student repeats the experiment. He keeps all the conditions the same, but uses the same starting mass of **powdered** calcium carbonate. On the grid, draw the line that the student should obtain for this experiment.

> You do not need to plot individual points for this line.

     Label this line **C**.   **(2 marks)**

Heat energy changes

1 Breaking bonds and making bonds involves energy transfers. Which row (**A**, **B**, **C** or **D**) in the table correctly describes these processes?

		Bond breaking	Bond making
☐	**A**	exothermic	exothermic
☐	**B**	exothermic	endothermic
☐	**C**	endothermic	exothermic
☐	**D**	endothermic	endothermic

(1 mark)

Guided

2 Describe, in terms of energy transfers, the difference between an exothermic process and an endothermic process.

> Think about whether heat energy is taken in or given out in these processes.

In an exothermic change or reaction, heat energy is ....

but in an endothermic change or reaction, heat energy is .... **(2 marks)**

3 Changes in heat energy occur when salts dissolve in water. They also occur in precipitation reactions.

(a) Magnesium nitrate solution reacts with sodium carbonate solution. Sodium nitrate solution and a precipitate of magnesium carbonate form. The temperature of the reaction mixture decreases.

State whether the reaction is exothermic or endothermic.

.... **(1 mark)**

(b) Give two types of reaction, which take place in aqueous solution, that are always exothermic.

....

.... **(2 marks)**

4 Magnesium reacts with dilute hydrochloric acid, forming magnesium chloride solution and hydrogen gas.

> You should be able to recall the formulae of elements and simple compounds.

(a) Write the balanced equation, including the state symbols, for this reaction.

.... **(3 marks)**

(b) Describe the measurements you would take to confirm that the reaction is exothermic.

> Outline what you would measure, the measuring apparatus and how you would use the results

....

....

.... **(3 marks)**

(c) Explain, in terms of breaking bonds and making bonds, why this reaction is exothermic.

....

....

.... **(3 marks)**

Reaction profiles

1 Methane burns completely in oxygen to form carbon dioxide and water vapour:

$$CH_4(g) + 2O_2(g) \rightarrow CO_2(g) + 2H_2O(g)$$
reactants   products

The diagram shows a simple reaction profile for this reaction.

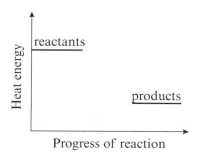

Explain, using information in the diagram, how you can tell that this reaction is exothermic.

Is heat energy given out or taken in during this reaction?

....

.... **(2 marks)**

2 Carbon burns completely in oxygen to form carbon dioxide: $C(s) + O_2(g) \rightarrow CO_2(g)$

(a) Complete the reaction profile for this reaction by showing the activation energy. **(2 marks)**

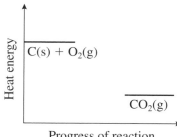

 Guided

(b) Give the meaning of the term **activation energy**.

the minimum .... **(1 mark)**

3 Calcium carbonate decomposes when heated strongly, forming calcium oxide and carbon dioxide:

$$CaCO_3(s) \rightarrow CaO(s) + CO_2(g)$$

The reaction is endothermic. Draw a reaction profile for the reaction, including the activation energy.

Label clearly the overall energy change in the reaction, and the activation energy. **(4 marks)**

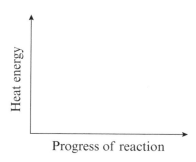

Calculating energy changes

1 Hydrogen and chlorine react together to form hydrogen chloride. The reaction can be modelled using the structures of the molecules involved:

$$H–H + Cl–Cl \rightarrow 2(H–Cl)$$

The table shows the bond energies for the bonds present in the reactants and products.

Bond	H–H	Cl–Cl	H–Cl
Bond energy (kJ mol⁻¹)	436	243	432

> **Guided**

(a) Calculate the energy taken in when the bonds in the reactants are broken.

$(1 \times 436) + (1 \times 243) =$ ...679...... kJ mol⁻¹   **(1 mark)**

> **Guided**

(b) Calculate the energy given out when the bonds in the products are formed.

$(2 \times 432) =$ ...864.......................... kJ mol⁻¹   **(1 mark)**

> **Guided**

(c) Use your answers to (a) and (b) to calculate the energy change in the reaction.

energy change = (energy in) − (energy out) = ...679 − 864 = −185....

.... kJ mol⁻¹   **(2 marks)**

(d) Explain, using your answer to (c), whether the reaction is exothermic or endothermic.

> The energy change is negative for an exothermic reaction.

...........exothermic as it was a negative energy........

...change..   **(2 marks)**

2 In the Haber process, nitrogen and hydrogen react together to produce ammonia. The reaction can be modelled using the structures of the molecules involved:

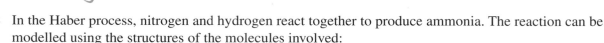

The table shows the bond energies for the bonds present in the reactants and products.

Bond	N≡N	H–H	N–H
Bond energy (kJ mol⁻¹)	945	436	391

Calculate the energy change for the Haber process.

> Remember to include **all** the N–H bonds present.

$(1 \times 945) + (3 \times 436)$

$945 + 1308$

$= 2253 \text{ kJ mol}^{-1}$

$(6 \times 391) = 2346$

$2253 - 2346$

$= -93$

...−93........... kJ mol⁻¹   **(4 marks)**

Crude oil

1 Crude oil is described as mainly a complex mixture of:

☐ **A** hydrogen and carbon   ☐ **C** polymers

☐ **B** alkenes   ☑ **D** hydrocarbons   **(1 mark)**

2 Crude oil is a **finite** resource. Explain what this means.

It has a limited supply It isnt
unlimited   **(2 marks)**

3 The diagram shows the structures of hexane and cyclohexane, two substances found in crude oil.

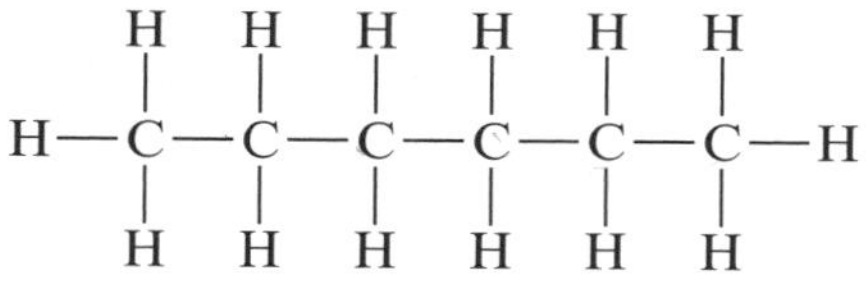

hexane 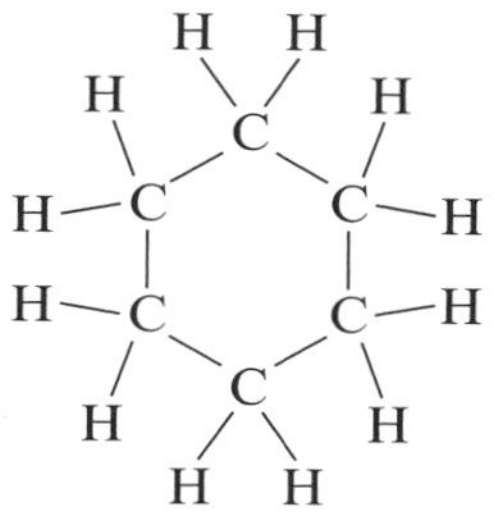   cyclohexane

(a) Write the molecular formula for hexane.

C₆H₄   **(1 mark)**

Guided

(b) Explain why hexane and cyclohexane are hydrocarbons.

They are compounds of hydrocarbon as it only
contains hydrogen and carbon   **(2 marks)**

4 Crude oil is an important source of fuels. Octane, C_8H_{18}, is a substance obtained from crude oil that is used as a fuel.

(a) Write a balanced equation for the complete combustion of octane in oxygen.

> The only products are carbon dioxide and water.

$C8H + 18 + O_2 = CO_2 + H_2O$ **(2 marks)**

(b) Describe the chemical test for carbon dioxide.

> State what you would do and what you would observe.

So closely Water las   **(2 marks)**

5 Crude oil is an important feedstock for the petrochemical industry. For example, poly(chloroethene) or PVC is a polymer made from chloroethene. The chloroethene itself is made in two stages from ethene, a substance produced from crude oil.

Explain the meaning of the term **feedstock**.

raw material to supply push   **(2 marks)**

Fractional distillation

1 Crude oil is separated into simpler, more useful mixtures by fractional distillation. The diagram shows the main fractions obtained from crude oil.

→ gases

→ petrol

→ kerosene

→ diesel oil

→ fuel oil

crude oil →

→ bitumen

(a) How do the fractions differ from **bottom** to **top** in the column?

☒ **A** The numbers of carbon atoms and hydrogen atoms in their molecules increase.

☐ **B** The boiling point of the fractions increases.

☐ **C** The viscosity of the fractions increases.

☐ **D** The ease of ignition increases.   **(1 mark)**

(b) Name the oil fraction that is used:

Use the diagram to help you.

(i) to surface roads and roofs

.... bitumen ....   **(1 mark)**

(ii) as a fuel for aircraft.

.... fuel oil Kerosene ....   **(1 mark)**

(c) Name two oil fractions that are used as fuels for cars.

.... diesel oil, petrol ....   **(2 marks)**

2 Most of the hydrocarbons in crude oil are members of a particular homologous series. Name this series.

.... Alkane homologous series ....   **(1 mark)**

3 Explain, in terms of its physical properties, why fuel oil is **not** a suitable fuel for cars.

Guided

The viscosity of fuel oil is too ..... thick ....

and it does not vaporise easily because its ..... boiling points much hig...   **(2 marks)**

4 Explain why the fraction leaving the top of the fractionation column has a low boiling point.

In your answer, name the bonds or forces overcome during boiling.

..... The gases donot conденs and leaves
..... at the top.x There are Weak forces between
..... molecules ....   **(3 marks)**

5 Describe how crude oil is separated using fractional distillation.

..... Crude oil Can be Separated by fractional
..... distillation as the hydrocarbons all have different
..... boiling points where it is evaporated the gases do
..... not Condense and leaves at the top ....   **(4 marks)**

Alkanes

1 Natural gas is a hydrocarbon fuel. It is mainly methane, CH_4. Which of the following substances cannot be released when methane burns in air?

☐ **A** water

☐ **B** carbon

☐ **C** hydrogen

☐ **D** carbon dioxide

(1 mark)

2 The alkanes form an homologous series of hydrocarbons.

> **Guided**

(a) State the general formula for the alkanes.

C_nH .... **(1 mark)**

(b) Dodecane is an alkane that has 12 carbon atoms in its molecules.

(i) Predict the molecular formula of dodecane.

.... **(1 mark)**

(ii) Name the products of complete combustion of dodecane.

> These products will be the same, whichever alkane undergoes complete combustion.

.... **(2 marks)**

(iii) State why complete combustion of alkanes involves oxidation.

.... **(1 mark)**

3 Similar to the alkanes, the alcohols form an homologous series of compounds. The table shows information about the first three members of the alcohol homologous series.

Name of alcohol	Molecular formula	Structure
methanol	CH_4O	H—C—O—H (with H above and below C)
ethanol	C_2H_6O	H—C—C—O—H (with H above and below each C)
propanol	C_3H_8O	H—C—C—C—O—H (with H above and below each C)

(a) Give two ways in which the molecules of methanol, ethanol and propanol are similar to each other.

....

.... **(2 marks)**

(b) State how the molecular formula of an alcohol differs from its neighbouring compounds.

> Compare the three molecular formulae. It may also help to compare the three structures.

.... **(1 mark)**

157

Incomplete combustion

1   Petrol is a hydrocarbon fuel. When it burns in air, waste products form and energy is transferred to the surroundings. Which row in the table correctly shows differences when 1 dm³ of petrol undergoes complete combustion or incomplete combustion?

	Complete combustion	Incomplete combustion
☐ **A**	more energy transferred	water vapour produced
☐ **B**	more energy transferred	water vapour not produced
☐ **C**	water vapour produced	more energy transferred
☐ **D**	water vapour not produced	more energy transferred

(1 mark)

2   When diesel oil burns in a limited supply of air, carbon monoxide gas and carbon particles are produced.

 Guided

(a) Explain why carbon monoxide is toxic.

When breathed in, carbon monoxide combines with ....

so .... **(2 marks)**

(b) Give a reason that explains why carbon particles may be harmful to health if breathed in.

....
(1 mark)

3   The diagram shows a bird's nest blocking the pipes leading to and from a central heating boiler.

Explain how the nest affects the safety of the people who live in the house.

....

....

....

....
(3 marks)

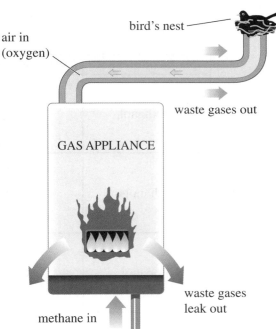

bird's nest

air in (oxygen)

waste gases out

GAS APPLIANCE

waste gases leak out

methane in

4   Propane, C_3H_8, is used in camping gas cylinders.

(a) Write a balanced equation for the complete combustion of propane.

Remember that there are two products when hydrocarbons burn completely in air.

.... **(2 marks)**

Guided

(b) Write a balanced equation for the incomplete combustion of propane, where equal amounts of carbon, carbon monoxide and carbon dioxide form.

C_3H_8 + ...O_2 → ....H_2O + ... + ... + ... **(2 marks)**

Acid rain

1 Nitrogen and oxygen can react together to form nitrogen dioxide, NO_2, which is a pollutant gas.

(a) Write a balanced equation for this reaction.

....

> Remember that elements that are gases at room temperature (apart from the noble gases) exist as diatomic molecules, X_2.

.... **(2 marks)**

> **Guided**

(b) Explain why oxides of nitrogen, such as nitrogen dioxide, are produced by working engines.

Oxygen and nitrogen from ....

react together at the high .... **(2 marks)**

(c) Many fuels used in engines contain impurities that cause the formation of another pollutant gas, which is acidic.

(i) Name the element present in these impurities that produces this gas.

.... **(1 mark)**

(ii) Describe, with the help of a balanced equation, how the element named in part (**i**) causes the formation of this gas.

....

.... **(2 mark)**

2 Sulfur dioxide dissolves in rainwater to produce an acidic solution. Write a balanced equation to show the reaction that forms dilute sulfurous acid, H_2SO_3. Include state symbols in your answer.

> **Guided**

$SO_2(........) + ...(........) \rightarrow ....(........)$ **(2 marks)**

3 Acid rain forms when sulfur dioxide dissolves in rainwater.

(a) The diagrams show two old gravestones. The one on the left is made from marble, and the other is made from granite. Explain why they are evidence for acid rain in the area.

> Marble contains calcium carbonate, $CaCO_3$, but granite does not.

IN
LOVING MEMORY OF
OUR MOTHER
MOLLIE CLARE LYNCH
DIED 10TH JANUARY 1900
AGED 80 YEARS
REST IN PEACE

IN
MEMORY OF
OUR BELOVED FATHER
WILLIAM TELFORD
DIED 16TH DECEMBER 1983
AGED 67 YEARS
REST IN PEACE

....

.... **(2 marks)**

(b) Describe two problems, other than the one shown in part (**a**), caused by acid rain.

....

.... **(2 marks)**

Choosing fuels

1 Crude oil and natural gas are finite resources because they take a very long time to form, or are no longer being made. Methane is a non-renewable fossil fuel that is found in natural gas.

> Non-renewable and finite have different meanings. Do not answer by writing 'it is not renewable'.

State why methane is described as **non-renewable**.

.... **(1 mark)**

2 Petrol, kerosene and diesel oil are fossil fuels.

(a) State the name of the substance from which these fuels are obtained.

.... **(1 mark)**

> Guided

(b) Give one example of how each fuel is used.

Petrol is used as a fuel for cars. Kerosene is used as a fuel for.................................

.... **(3 marks)**

3 Hydrogen and petrol may both be used as fuels for cars.

(a) Write the balanced equation for the reaction between hydrogen and oxygen.

> There is only one product.

.... **(2 marks)**

(b) Petrol is a complex mixture of hydrocarbons. Name one product of the complete combustion of petrol that is **not** produced when hydrogen burns.

.... **(1 mark)**

4 The table shows some information about hydrogen and petrol.

Fuel	Energy released by 1 kg of fuel (MJ)	State at room temperature and pressure	Volume of 1 kg at room temperature and pressure (dm^3)
hydrogen	141.8	gas	12 000
petrol	47.3	liquid	1.36

> Guided

(a) Calculate, to 3 significant figures, the volume of hydrogen needed to release 100 MJ of energy.

$$\text{volume needed} = 12\,000 \times \left(\frac{100}{141.8}\right) = \text{.................................} \ dm^3 \quad \textbf{(1 mark)}$$

(b) Calculate, to three significant figures, the volume of petrol needed to release 100 MJ of energy.

.... **(1 mark)**

(c) Using information in the table and your answers to (**a**) and (**b**):

(i) identify an advantage of using hydrogen rather than petrol as a fuel for cars.

.... **(1 mark)**

(ii) identify an advantage of using petrol rather than hydrogen as a fuel for cars.

.... **(1 mark)**

Cracking

1   Alkanes and alkenes form two different homologous series. Which row in the table correctly describes each type of hydrocarbon?

	Alkanes	Alkenes
☐ **A**	saturated	contain only C–H and C–C bonds
☐ **B**	unsaturated	contain only C–H and C–C bonds
☐ **C**	contain only C–H and C–C bonds	saturated
☐ **D**	contain only C–H and C–C bonds	unsaturated

(1 mark)

2   In the diagram below, a cracking reaction is modelled using the structures of the molecules.

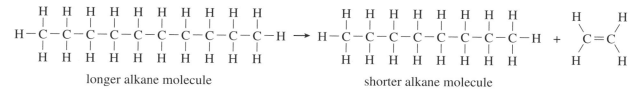

            longer alkane molecule   shorter alkane molecule

  (a)   Name the **type** of hydrocarbon shown by the smallest molecule above.

.... **(1 mark)**

  (b)   Write the balanced equation, using molecular formulae, for this cracking reaction.

> Count the carbon atoms and hydrogen atoms in each molecule to work out the formulae needed.

.... **(2 marks)**

3   Explain what is meant by **cracking**.

 Guided

a reaction in which larger alkanes are broken down into ....

.... **(2 marks)**

4   Crude oil is separated into more useful mixtures called fractions by fractional distillation.

  (a)   Explain why an oil refinery may crack the fractions containing larger alkanes.

....

.... **(2 marks)**

  (b)   The graphs show the composition of crude oil obtained from two different oil wells.

        Explain why oil from one of the oil wells is probably **not** sent for cracking.

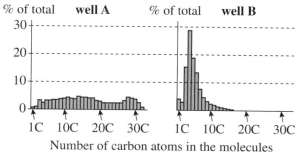

....

.... **(2 marks)**

Extended response – Fuels

Camping gas is a mixture of propane and butane, obtained from crude oil. It is a rainy day and some campers are making tea inside their tent. Incomplete combustion of the camping gas could occur if the campers do not take adequate precautions.

Explain how incomplete combustion of hydrocarbons such as propane and butane occurs, and the problems that it can cause in a situation similar to this one. You may include a balanced equation in your answer.

> You will be more successful in extended response questions if you plan your answer before you start writing. Think about the following questions to help you:
>
> How does incomplete combustion occur?
>
> Which is a more efficient use of fuels, complete or incomplete combustion, and why does this matter?
>
> What are the products of the incomplete combustion of hydrocarbons? What problems do they cause?
>
> Remember that carbon dioxide may be produced during incomplete combustion as well as during complete combustion, and so does not explain the problems that incomplete combustion causes.

....

....

....

....

....

....

....

....

....

.... **(6 marks)**

> Continue your answer on your own paper. You should aim to write about half a side of A4.

The early atmosphere

1 From where did the gases that formed the Earth's earliest atmosphere come?

☐ **A** combustion   ☐ **C** photosynthesis

☑ **B** volcanic activity   ☐ **D** condensation

(1 mark)

2 The table shows possible percentages of four gases in the Earth's early atmosphere.

Name of gas	Percentage in atmosphere
nitrogen	trace amount
oxygen	trace amount
water vapour	75
carbon dioxide	14

(a) Name the most abundant gas in the modern atmosphere.

> It is one of the four gases shown in the table.

.... *Nitrogen* .... **(1 mark)**

(b) Explain how the Earth's oceans formed.

.... *carbon dioxide dissolved into the* .... *oceans* .... **(2 marks)**

(c) The Earth's atmosphere today contains about 0.04% carbon dioxide. Explain how the oceans contributed to the change in the percentage of carbon dioxide in the atmosphere.

.... *The oceans dissolve carbon dioxide* ....

.... **(2 marks)**

3 Describe the chemical test for oxygen.

> Write down what you would do and what you would observe.

.... *You light a splint you blow it until there* .... *is a glowing lighter then put it in a test tube* .... **(2 marks)** *if it re-lights oxygen is present*

4 The Earth's atmosphere today contains more oxygen than its early atmosphere did.

> In your answer, include the name of the process involved.

> **Guided**

Explain why the percentage of oxygen in the atmosphere has gradually increased.

The growth of primitive plants ...... *is because photosynthesis begins* ...... *and oxygen builds up in the Ocean* ......

.... **(3 marks)**

5 Different teams of scientists may give different percentages for the various gases in the Earth's early atmosphere. Suggest a reason that explains why this happens.

.... *This could be because most of the carbon* .... **(1 mark)** *dioxide is dissolved in the ocean* ✗ *No humans were on the earth*

Greenhouse effect

1 Carbon dioxide and water vapour are described as greenhouse gases.

(a) Name another greenhouse gas, often released as a result of livestock farming.

.... **(1 mark)**

(b) The use of fossil fuels increases the concentration of carbon dioxide in the atmosphere.

> What type of substance is found in fossil fuels such as petrol? What happens when they are burned?

(i) Explain why the use of fossil fuels causes the release of carbon dioxide.

....

.... **(2 marks)**

(ii) Explain why the release of carbon dioxide from recent human activities, such as using fuels, is causing an increase in the concentration of carbon dioxide in the atmosphere.

> What processes release carbon dioxide to the atmosphere and remove it, and how fast do they work?

....

....

.... **(3 marks)**

2 Explain what is meant by the **greenhouse effect**.

> **Guided**

Various gases in the atmosphere, such as carbon dioxide, absorb ....

.... and then release ....

which .... **(3 marks)**

3 The graph shows how the mean global temperature, and the percentage of carbon dioxide in the atmosphere, have changed over the last 220 000 years.

(a) Suggest a reason that explains why these measurements may not be certain.

——— difference in temperature

▬▬ percentage of CO_2 in the air

Thousands of years before today

....

.... **(1 mark)**

(b) Describe the relationship between the carbon dioxide level and global temperature, shown in the graph.

....

.... **(2 marks)**

(c) State an environmental effect of increasing mean global temperatures.

.... **(1 mark)**

Extended response – Atmospheric science

The graph shows the change in mean global temperature, and the concentration of carbon dioxide in the atmosphere, between the years 1850 and 2005.

> A higher temperature than the mean temperature gives a positive temperature change on the graph.

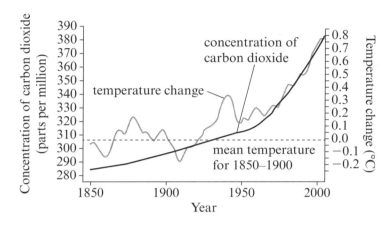

Evaluate whether these graphs provide evidence that human activity is causing the Earth's temperature to increase. In your answer, explain how carbon dioxide acts as a greenhouse gas, and describe processes that release or remove carbon dioxide.

> You will be more successful in extended response questions if you plan your answer before you start writing.
>
> You should be able to evaluate evidence for human activity causing climate change. This could include correlations between the change in atmospheric carbon dioxide concentration, the consumption of fossil fuels and temperature change.

....

....

....

....

....

....

....

....

....

....

....

....

.... **(6 marks)**

Key concepts

1　Complete the table for units of physical quantities and their abbreviations.

Guided

| ampere | A | joule | J | pascal | Pa | coulomb | C |
| mole | Mol | watt | W | newton | N | ohm | Ω |

(2 marks)

2　Explain the difference between a base unit and a derived unit.

Base units are defined by a particular process of measuring a base quantity whereas derived units are algbraic combination

(2 marks)

3　Convert each quantity.

(a)　750 grams to kilograms

Maths skills　1000 g = 1 kg; 1000 W = 1 kW; 60 s = 1 minute; 1000 mm = 1 m; 1 000 000 J = 1 MJ

.... 0.75 .... kg　**(1 mark)**

(b)　0.75 kilowatts to watts

.... 750 .... W　**(1 mark)**

(c)　25 minutes to seconds

.... 1500 .... s　**(1 mark)**

(d)　30 millimetres to metres

.... 0.03 .... m　**(1 mark)**

(e)　3 megajoules to joules

.... 3 000 000 .... J　**(1 mark)**

4　Write each quantity in the unit shown and then in standard form.

Guided

(a)　frequency of 2.5 kHz

Hz: 2.5 kHz = 2500 Hz

standard form: 2.5×10^3 Hz　**(2 marks)**

(b)　length of 8 nm

m: .. 8000 ..

standard form: .. 8×10^3 ..　**(2 marks)**

5　Calculate the speed of a car that takes 10.5 s to travel 75 m. Give your answer to 5 significant figures.

> Look at the number that follows the significant figure you are asked to consider (in this case the 5th one). If it is greater than 5, round the 5th figure up, if it is less then round down, e.g. 1.23076923 would become 1.2308 to 5 significant figures.

speed = ... 7.1 .... m/s　**(2 marks)**

Scalars and vectors

1 (a) Write each quantity in the correct part of the table.

acceleration displacement ~~speed~~ ~~energy~~
~~temperature~~ ~~mass~~ force velocity
momentum ~~distance~~

> A scalar has only a magnitude (size) but a vector has both a magnitude **and** a direction.

Scalars	Vectors
Mass Speed displane energy Temperature	Momentum Acceleration displacment force Velocity

(2 marks)

(b) (i) Give one example of a scalar from your table and explain why it is a scalar.

.......Speed............... is a scalar because ......It hase magnitude
.....and...... but no Specific direction...... **(2 marks)**

(ii) Give one example of a vector from your table and explain why it is a vector.

.......force......... is a vector because ......because it has size
and ...a direction............................. **(2 marks)**

2 At the swimming pool, two swimmers are practising for a swimming gala. They swim from opposite ends of the pool. The first swimmer dives in from the left side and swims the length of the pool at a velocity of 1.3 m/s. The second swimmer then swims from the right at a velocity of −1.4 m/s.

(a) Explain why the velocity is used in this example instead of the speed.

.....They are going in a direction.............
.... **(2 marks)**

(b) Explain why the second swimmer's velocity has a negative value.

.....They take to the right as negative.... **(1 mark)**

3 (a) Which of the following is not a scalar?

☐ **A** energy ☐ **C** mass

☐ **B** temperature ☑ **D** weight **(1 mark)**

(b) Give a reason for your answer to (a).

....Weight has a direction.....................
.... **(1 mark)**

4 An aeroplane flies in a straight line between two airports. The pilot knows that there will be a strong wind blowing at an angle of 60° to the direction in which the aeroplane will be travelling. Explain why it is important that the pilot uses vectors when planning the route the aeroplane will take.

.....The pilot heads to make sure the
.....Wind doesnt blow of track........
.... **(3 marks)**

Speed, distance and time

1 The distance/time graph shows a runner's journey from his home to the park.

(a) State the letter that corresponds to the part of the runner's journey where he:

    (i) stops   ....*B*............   **(1 mark)**

    (ii) runs fastest.   ....*C*............   **(1 mark)**

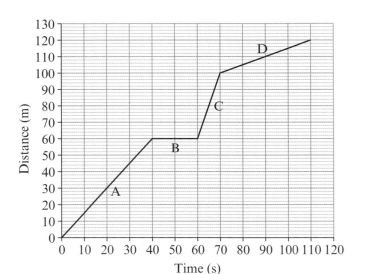

Guided

(b) Calculate the runner's speed in part A of his journey.

In part A, he travels ..*60*.... m in ..*40*.... s.

speed = distance ÷ ......*Time*..........

                          speed = ...*1.5*.............. m/s **(3 marks)**

(c) When the runner arrives at the park his displacement from home is less than the distance he has travelled. Explain this difference.

.........*displacement is the Shortest route possible from the*.........

.........*Start while distance is The ranke he has took*......... **(2 marks)**

2 The lift in a wind turbine tower takes 24 s to go from the ground to the generator 84 m above.

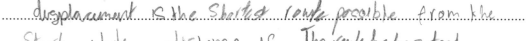
Speed = distance ÷ time only.
Velocity is speed in a given direction.
For example, speed = 20 m/s but velocity = 20 m/s East

(a) Calculate the speed of the lift. State the unit.

               speed = ..*3.5*........... unit *m/s*........... **(3 marks)**

(b) State the velocity of the lift.

..........................*3.5 M/S North*   *✗ 3.5 m/s wp*.............. **(1 mark)**

3 An athlete runs at a constant speed of 5 m/s around a running track. A complete lap is 400 m.

Calculate the time it takes for the athlete to complete one lap.

                               time = ...*80*.......... s **(2 marks)**

Equations of motion

1 Draw a line from each symbol to its correct description. One has been done for you.

Initial velocity means the velocity when time = 0.

Symbol		Description
v		acceleration
u		time
a		distance
x		final velocity
t		initial velocity

(2 marks)

2 (a) A racing car takes 8 seconds to speed up from 15 m/s to 25 m/s. Calculate its acceleration.

You may find the equation
$a = (v - u) \div t$ useful.

$$(25 - 15) \div 8$$

acceleration = ...1.25...... m/s² **(3 marks)**

(b) The racing car now accelerates at the same rate for 12 seconds, from 25 m/s to a higher velocity. It travels 300 m during this time. Calculate its final velocity.

You may find the equation
$v^2 - u^2 = 2 \times a \times x$ useful.

$$v^2 = u^2 + 2 \times a \times (x) = 25\,(m/s) + 2\,(1.25\,m/s^2 \times 300\,m)$$
$$= 1375 \quad V = \sqrt{1375} = 37\,m/s$$

velocity = ...1500 37 m/s...... m/s **(3 marks)**

(c) The car now slows down to 5 m/s from the velocity calculated in (b) at a rate of −2 m/s².
Calculate how far the car travels when decelerating to this new final velocity.

$$v^2 - u^2 = 2\,a \times (x) \quad \text{So} \; x = v^2 - u^2 \div 2 \times a\,(1) = 5(m/s)$$
$$\frac{1375\,(m^2/s^2)(2 \times -2\,m/s^2) = -1350\,m^2/s^2 \div -4\,m/s^2 =}{}$$

337.5

distance = ...337.5 x 337.5...... m **(3 marks)**

169

Velocity/time graphs

1 A cyclist takes 5 seconds to reach maximum velocity of 4 m/s, from being stationary, moving in a straight line.

(a) Calculate the cyclist's acceleration.

$$\frac{5 \times 4}{2} = 10$$

acceleration = ...10.................unit .m/s.c...... **(3 marks)**

(b) The cyclist travels at constant velocity for 15 seconds and then takes another 15 seconds to slow down to a stop. Explain how the total distance travelled could be calculated by drawing a graph of the ride.

....

.... **(3 marks)**

2 The velocity/time graph shows how the velocity of a car changes with time.

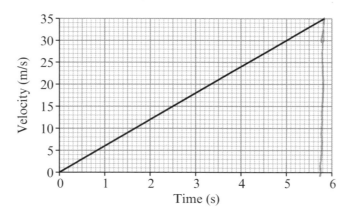

(a) This graph can be used to analyse the car's journey. Choose **two** correct statements that describe the information the graph shows.

☒ **A** the distance the car travelled   ☐ **C** the acceleration of the car

☐ **B** how long the car was stopped   ☒ **D** the constant velocity of the car

(2 marks)

(b) Draw a triangle on the graph to show the acceleration and the time taken.   **(1 mark)**

⟩ **Guided** ⟩   (c) Calculate the acceleration of the car.

change in velocity = ... m/s, time taken for the change = ... s

$$\text{acceleration} = \frac{\text{change in velocity}}{\text{time taken}} = $$

acceleration = ... m/s^2 **(2 marks)**

(d) Use the graph to calculate the distance travelled by the car in the first 5 s.

> Work out the area under the graph.

distance = ...m **(2 marks)**

Determining speed

1   Draw a line from each activity to its correct speed. One has been done for you.

Activity
commuter train
running
speed of sound in air
walking

Speed
330 m/s
1.5 m/s
3.0 m/s
55 m/s

(1 mark)

2   The diagram shows a light gate being used to measure the speed of a model vehicle.

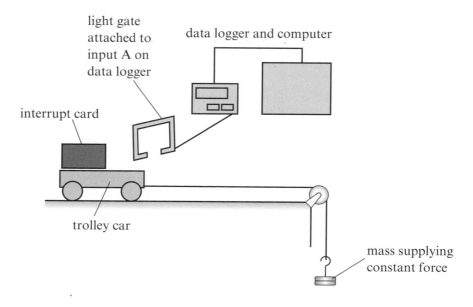

light gate
attached to
input A on
data logger

data logger and computer

interrupt card

trolley car

mass supplying
constant force

Guided   (a)   Describe why a card is fixed to the vehicle in this experiment.

The light beam is ....

as it enters the light gate and this starts the timer. When the card has passed

through, and ....

.... **(3 marks)**

(b)   State how the speed is found using this method.

.... **(1 mark)**

3   State two reasons why using light gates and a computer may be a more reliable method than using a person with a stop watch and a ruler to measure the speed of a toy car.

> **Practical skills**   Light gates can measure instantly so computers can calculate speeds over very short distances. Consider this advantage over measurement by a person.

....

....

.... **(3 marks)**

Newton's first law

1 A submarine is travelling at a constant depth in the sea. It starts to move forwards. Draw a free-body force diagram for all the forces acting on the submarine. Label these forces. **(2 marks)**

2

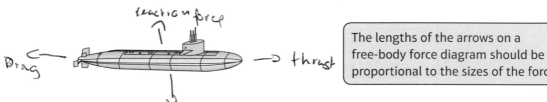

reaction force

Drag ← → thrust

weight

> The lengths of the arrows on a free-body force diagram should be proportional to the sizes of the forces.

2 A speed skater is standing on the ice waiting for the start of a race.

Guided

(a) Describe the action and reaction forces acting on the skater and her skates.

The action is the ...downward force on the Skater... and the

reaction is ......upwards force...... **(2 marks)**

(b) The race begins and the skater pushes against the ice producing a forward thrust on the skates of 30 N. There is resistance from the air of 10 N and friction on the blades of 1 N. Calculate the resultant force.

> Add up all the forces in a straight line. Give forces that act opposite to the thrust a minus sign.

30 − 10 − 1 = 19

force = ...19...... N **(2 marks)**

2

(c) During the race the resistive forces become equal to the forward thrust. Explain what happens to the velocity of the skater.

......The forces are balanced. The Skater will move at a......

......Constant Speed.............. **(2 marks)** 2

(d) At the end of the race the skater stops skating. Explain what happens next before the skater comes to a halt.

......The Skater will Speed but the resultant force will......

......be more So he will stop........ **(2 marks)** 2

3 A space probe falls towards the Moon. In the Moon's gravitational field the probe has a weight of 1700 N. The probe fires rockets giving an upward thrust of 1900 N.

(a) Calculate the resultant force on the space probe.

1900 − 1700 = 200

2

resultant force = ... N **(2 marks)**

(b) Explain the changes in the probe's velocity.

......The force in the opposite direction to the movent......

......will Slow it down it can reverse its motion...... **(2 marks)** 2

Newton's second law

1   In an experiment a student pulls a force meter attached to a trolley along a bench. The trolley has frictionless wheels. The force meter gives a reading of 5 N.

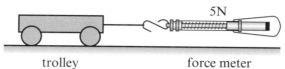

trolley   force meter

Guided   (a) Describe what happens to the trolley.

The trolley will ....

in the direction.. **(2 marks)**

Guided   (b) The student stacks some masses on the trolley and again pulls it with a force of 5 N. Explain why the trolley takes longer to travel the length of the bench.

The acceleration is ....because... **(2 marks)**

2   When the Soyuz spacecraft returns to Earth from the International Space Station it is slowed by friction with the air. The spacecraft has a mass of 3000 kg and the craft slows with an average acceleration of −13 m/s².

(a) Calculate the average resultant force acting on the spacecraft. State the unit.

force = ... unit .... **(3 marks)**

(b) State the direction in which the force acts.

.... **(1 mark)**

3   A Formula One racing car has a mass of 640 kg. A resultant force of 10 500 N acts on the car.

(a) Calculate the acceleration on the racing car. State the unit.

> Newton's second law is $F = m \times a$
> where F = (unbalanced) force,
> m = mass and a = acceleration

acceleration = ... unit .... **(3 marks)**

(b) Explain what will happen to the acceleration of the car as its fuel tank empties, assuming the resultant force remains constant.

....

.... **(2 marks)**

 **Practical skills**

Weight and mass

 Guided

1   The lunar roving vehicle (LRV), driven by astronauts on the Moon, has a mass of 210 kg on Earth. State the mass of the unchanged LRV on the Moon. Give a reason for your answer.

The mass of the LRV on the Moon is .... kg

because ....

....

.... **(2 marks)**

2   Which of the following is **not** a description of weight?

☐ **A**   Weight is a type of force.

☐ **B**   Weight is measured in kilograms (kg).

☐ **C**   The weight of a mass changes according to gravitational field strength.

☐ **D**   Weight is measured in newtons (N).   **(1 mark)**

3   Calculate the total weight of a backpack of mass 1 kg, containing books with a mass of 2 kg and trainers with a mass of 1.5 kg. Take gravitational field strength (g) to be 10 N/kg.

> Use the equation relating weight to mass and gravitational field strength.

weight = ... N **(3 marks)**

4   Kate is about to fly to Europe on holiday. The total baggage allowance is 20 kg. Kate only has scales that weigh in newtons. Determine the items that Kate can take on holiday, as well as her clothes, to get the mass as close as possible to the baggage allowance. Show your calculations. Take gravitational field strength (g) to be 10 N/kg.

| laptop 45 N | camera bag 55 N | walking boots 25 N | jacket 35 N | clothes 105 N |

total baggage = ... kg **(3 marks)**

🧪 Practical skills **Force and acceleration**

A ramp, a trolley, masses and electronic light gates can be used to investigate the relationship between force, mass and acceleration.

1 State one advantage of using electronic measuring equipment to determine acceleration compared to using a ruler and stopwatch.

....

....

.... **(2 marks)**

2 Describe the relationship between acceleration and mass.

.... **(1 mark)**

3 Explain why it is necessary to use two light gates when measuring acceleration in this experiment.

> **Guided**

Acceleration is calculated by the change in speed ÷ time taken, so ....

....

.... **(2 marks)**

4 (a) Describe the conclusion that can be drawn from this experiment.

> **Guided**

For a constant slope...

....

.... **(2 marks)**

(b) Identify which of Newton's laws can be referred to in verifying the results of this experiment.

> The quantities of force, mass and acceleration are linked in this equation.

.... **(1 mark)**

5 Suggest one hazard associated with this experiment and two safety precautions that could be taken to minimise the risk of harm to the scientist.

> Consider the potential dangers of using accelerated masses or electrical equipment.

....

....

....

....

....

.... **(3 marks)**

Circular motion

1 (a) Explain why the velocity of a satellite is constantly changing even though its speed remains constant.

The velocity of an orbiting satellite changes because ....

....

even though ....

.... **(2 marks)**

(b) Explain why the Moon can be described as accelerating in its orbit round the Earth.

> Refer to forces in your answer.

....

.... **(1 mark)**

2 What is the name of the force that is at 90° to the motion of a satellite?

☐ **A** acceleration

☐ **B** centripetal force

☐ **C** circular motion

☐ **D** orbiting force   **(1 mark)**

3 Name the force that acts as the centripetal force in each example.
Add another example of your own for each force.

Force	Example 1	Example 2
	a lasso used to catch cattle	
	Venus orbiting the Sun	
	a cyclist going around a velodrome track	

(6 marks)

4 A student presents his project on the moons of Jupiter and uses a ball tied to string to model their motion by rotating the ball around in a circle, holding on to the string.

> Consider how the forces act together.

(a) State the force that the string is modelling.

.... **(1 mark)**

(b) Explain why the ball is accelerating.

.... **(1 mark)**

Momentum and force

1 What is the momentum of a 10 000 kg lorry moving at 4 m/s?

    ☐ **A**   2500 kg m/s

    ☐ **B**   40 000 kg m/s

    ☐ **C**   14 000 kg m/s

    ☐ **D** 4×10^{-4} kg m/s   **(1 mark)**

2 (a) Explain how force is related to momentum.

    ....

    .... **(2 marks)**

   (b) A car with a mass of 1500 kg is travelling at 25 m/s along a motorway. It crashes into a central barrier and stops in 1.8 seconds resulting in a momentum of zero. Calculate the change in momentum of the car.

    change in momentum = ... kg m/s   **(3 marks)**

> **Guided**

   (c) Explain how a large force is exerted on a passenger in a vehicle in the event of a car crash and how this can be reduced.

    The forces exerted on the passenger are large when ....

    ....

    Fitting ....

    So this will reduce the .... **(4 marks)**

3 Calculate the force on a motorcycle of mass 500 kg as it speeds up from 10 m/s to 15 m/s in 20 s.

> You may find this equation useful:
> *change in momentum = resultant force × time*

    force = ... N   **(3 marks)**

4 Explain what a hockey player needs to consider when hitting the hockey ball with a hockey stick, to send the ball as far as possible down the pitch.

    ....

    ....

    .... **(3 marks)**

Newton's third law

1 Select the statement that summarises Newton's third law.

☐ **A** For every action there is a constant reaction.

☐ **B** The action and reaction forces are different due to friction.

☐ **C** Reaction forces may be stationary or at constant speed.

☐ **D** For every action there is an equal and opposite reaction. **(1 mark)**

Guided

2 Calculate the momentum of a car with a mass of 1200 kg moving at 30 m/s from north to south.

> You may find this equation useful:
> momentum = mass × velocity

momentum = ...

momentum = ... kg m/s in the .... direction. **(3 marks)**

3 Dima and Sam are driving dodgem cars at a funfair. The total mass of Dima and his car is 900 kg. He is moving west at 1.5 m/s.

(a) Calculate the momentum of Dima and his car.

momentum = ... kg m/s **(1 mark)**

(b) Sam and his car also have a total mass of 900 kg but his car is travelling faster than Dima's car, at 3 m/s west. Sam's car collides with the back of Dima's car and both cars move forward together.

(i) Calculate the momentum of Sam and his car just before the collision.

momentum = ... kg m/s **(1 mark)**

(ii) Explain what happens to the sum of the momentum of both cars after the collision.

....

.... **(2 marks)**

(iii) Calculate the velocity of both cars as they move off together after the collision.

velocity = ... m/s **(3 marks)**

4 A skater with a mass of 50 kg skates across the ice at 7.2 m/s in a straight line travelling north. She collides with her stationary partner who has a mass of 70 kg. They glide off together northwards. Calculate the velocity with which the pair glide across the ice.

velocity = ... m/s **(3 marks)**

Human reaction time

1 Reaction time is an important consideration in driving a vehicle safely. Which is the distance travelled due to the reaction time of a driver?

  ☐ **A**   overall stopping distance

  ☐ **B**   thinking distance

  ☐ **C**   braking distance

  ☐ **D**   reaction distance   **(1 mark)**

2 Explain how human reaction time is related to the brain.

>Guided

Human reaction time is the ....

....

It is related to .... **(2 marks)**

3 Explain how to measure human reaction times using a ruler.

....

....

....

.... **(3 marks)**

4 (a) State the range of reaction times of an average person to an external stimulus.

.... **(1 mark)**

  (b) Describe why people in certain professions train themselves to improve their reaction times. Give two examples and comment on why improved reaction times would be important in each case.

> Examples of professions you could use are driving instructors and helicopter pilots.

....

....

....

....

.... **(4 marks)**

5 A rabbit runs across the road 50 metres in front of a car. Calculate the reaction time of a driver who covers a distance of 25 metres travelling at a speed of 20 m/s between seeing the rabbit and putting his foot on the brake.

> You may find this equation useful: *speed = distance ÷ time*

....

....

reaction time .... s   **(2 marks)**

Stopping distances

1 (a) Identify the two factors that make up the overall stopping distances of a vehicle.

...... Thinking distance, braking distan ......

(2 marks)

(b) Write suitable headings for the table below.

> Look at your answers to **1(a)** and the example factors in the table to help you.

(1 mark)

> Guided

(c) Complete the table below to summarise the factors that affect overall stopping distance.

> Separate the factors that may affect the reaction time of a driver from those that affect the vehicle.

Factors increasing overall stopping distance	
1. Thinking distance will increase if	2. Stopping dch will increase if
driver being tired	the car's speed increases
the driver is distracted	Tyres are worn
Alchol in driver	Brakes are worn
+	Weight of car

(2 marks)

(d) Compare the overall stopping distance when a car's tyres are new and when they are worn.

...... The Stopping distance when a cart tyre ......
...... are new will be much shorter as there is more grp ...... **(2 marks)**

2 This poster highlights the dangers of using a mobile phone whilst driving.

State the component of overall stopping distance that using a mobile phone when driving could affect.

Give a reason for your answer.

...... This damages the thinking distance as you can't concentrate on the phone ......

You can't concentrate on the road and your mobile phone.

...... **(2 marks)**

3 Recent proposals have been made to increase the national speed limit in certain cases. Suggest how these proposals might increase the risk of damage to vehicles and their passengers.

...... The Stopping distance Increases at youll b faster ......

(3 marks)

Extended response – Motion and forces

A student investigates circular motion by tying a 57 g tennis ball to a string, which is then rotated in a horizontal plane at constant speed. The student counts the number of rotations.

Explain the principles being investigated in this experiment and identify important factors that influence the results.

> You will be more successful in extended response questions if you plan your answer before you start writing.
>
> The question asks you to give a detailed explanation of acceleration and centripetal force. Think about:
>
> - why the tennis ball is described as accelerating
> - how centripetal force is described
> - how the student can improve the investigation by changing variables
> - how the student can improve data collection
> - identify the importance of control variables.
>
> You should try to use the information given in the question.

....

....

....

....

....

....

....

....

....

....

....

....

....

.... **(6 marks)**

Energy stores and transfers

1 Which of the following is not an energy store?

☐ **A** chemical

☐ **B** light

☐ **C** thermal

☐ **D** kinetic

(1 mark)

2 Explain how an energy transfer diagram supports the law of conservation of energy.

The energy transfer diagram shows that ....

....

....

.... **(2 marks)**

3 A footballer has a breakfast of cereal and toast before setting off for a training session at the club. Complete the flow chart to show how energy is transferred to other stores.

> Write the correct store of energy in each space.

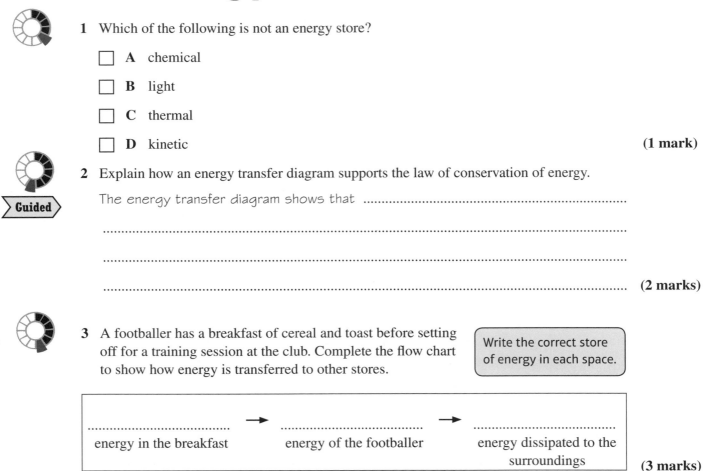

.... → .... → ....

energy in the breakfast   energy of the footballer   energy dissipated to the surroundings

(3 marks)

4 The bar graphs below illustrate energy stores before each energy transfer occurs. Add bars to the graphs to show energy stores for after each energy transfer has occurred.

(a) a bobsleigh at the top of a slope and halfway down the slope

(b) a petrol lawnmower before use and in use

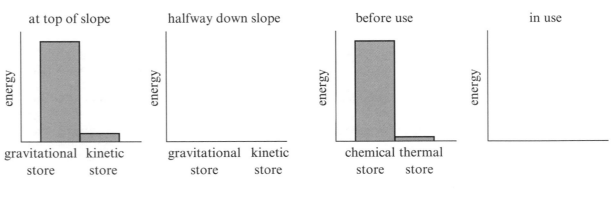

(3 marks)   **(4 marks)**

Efficient heat transfer

1   Identify the most suitable material, from the table below, for building an energy-efficient garage. Give a reason for your answer.

The larger the relative thermal conductivity, the more heat will be conducted through the material.

Material	Relative thermal conductivity
brick	1.06
concrete	1.00
sandstone	2.20
granite	2.75

....

.... **(2 marks)**

2   (a)   Some houses are built with very thick walls. Explain how these walls help to keep the houses warm in the winter in cold countries.

....

....

.... **(2 marks)**

(b)   In hot countries, such as Greece, traditional houses have thick walls with small windows. Explain why these houses in a hot country also have thick walls.

....

....

.... **(2 marks)**

3   A crane lifts a box to the top of a building. 1 000 000 joules is transferred to the gravitational store when the box is moved from the bottom to the top of the building. The crane uses fuel with 4 000 000 joules in a chemical store. Calculate the efficiency of the crane.

> **Guided**

useful energy transferred = energy transferred to the box = ...

total energy used by the crane = the energy stored in the fuel = ...

efficiency = ... **(2 marks)**

4   (a)   The motor in a food blender has an efficiency of 20%. The motor transfers 40 joules per second into the kinetic store. Calculate the energy that is transferred to the motor each second.

energy transferred each second = ... J   **(3 marks)**

(b)   State the power of the motor. Give the unit.

power = ... unit .... **(1 mark)**

Energy resources

1 Some of the sources of renewable energy listed below are only available at certain times, while other sources can be used at any time.

| hydroelectric | tidal | solar | wind | geothermal |

Guided (a) Name the sources of renewable energy in the list that are always available.

Hydroelectric and .... **(1 mark)**

Guided (b) Explain why it is an advantage to have a source of energy available at any time.

> Think about how the weather affects some renewable energy sources.

Demand is greatest ....

Demand may be high when .... **(2 marks)**

2 A hydroelectric power station is used to produce electricity when demand is high.

(a) Explain why the hydroelectric power station is a reliable producer of electricity.

....

.... **(2 marks)**

(b) Give one reason why we cannot use hydroelectric power stations in more places in the UK.

....

.... **(1 mark)**

3 Comment on each statement referring to the use of fossil fuels with regard to environmental impact.

> Think about the possible consequences of the statements describing the use of fossil fuels.

(a) Carbon dioxide is released as a result of burning fossil fuels.

....

.... **(2 marks)**

(b) Burning fossil fuels produces sulfur dioxide and nitrogen oxides.

....

.... **(2 marks)**

(c) Fossil fuel power stations can be built away from areas of natural beauty such as coasts, estuaries and mountains.

....

.... **(2 marks)**

4 Some people say that we have passed the time of 'peak oil'. After this time, the amount of crude oil extracted will decrease and prices for fuel will rise rapidly. Other people say that we will not pass this peak until 2020. Suggest why there is uncertainty about peak oil.

....

.... **(2 marks)**

Had a go ☐ Nearly there ☐ Nailed it! ☐

Patterns of energy use

1 The graphs show patterns of energy use and human population growth.

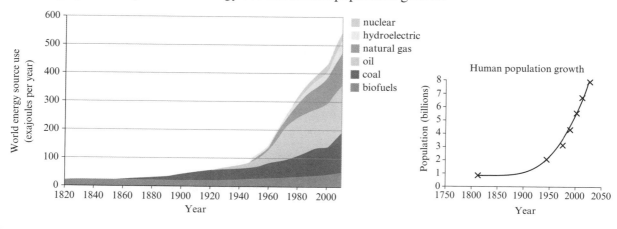

(a) Give three reasons why energy consumption rose significantly after the year 1900.

After 1900 the world's ....

There was development in ....

and .... **(3 marks)**

(b) (i) Identify which category of energy resources has been the main contributor to world energy consumption since the year 1900.

.... **(1 mark)**

(ii) Suggest two reasons why the consumption of energy resources has increased in the developed world.

....

.... **(2 marks)**

(iii) Suggest a reason why nuclear energy resources only appear after 1950.

.... **(1 mark)**

(iv) Identify a renewable resource from the graph that makes use of gravitational potential energy.

.... **(1 mark)**

2 If the patterns in energy consumption are similar to the patterns in the world's population growth, discuss the issues resulting from the continuing use of energy in the way shown in the graph in Q1.

> Consider finite non-renewable resources and increasing demand due to population, transport and industrial growth.

....

....

....

.... **(6 marks)**

Continue on a separate sheet if you need more space to complete your answer.

Potential and kinetic energy

1 Identify the correct equation for calculating gravitational potential energy.

☐ **A** $\Delta GPE = m \times v \times h$ ☐ **C** $\Delta GPE = m \times F \times a$

☐ **B** $\Delta GPE = \frac{1}{2} m \times v^2$ ☐ **D** $\Delta GPE = m \times g \times h$ **(1 mark)**

2 Calculate the kinetic energy of a cyclist and her bicycle, with combined mass of 70 kg, travelling at 6 m/s.

Guided

kinetic energy = $\frac{1}{2} mv^2$ = ...

kinetic energy = ... J **(2 marks)**

3 In the Middle Ages battering rams were used to smash down the doors of castles. The battering ram was a log of mass 2000 kg suspended by ropes. When the log was pulled back, it rose by 0.5 m.

0.5 m
upwards

(a) Calculate the gravitational potential energy gained by the log when it was pulled back. Assume $g = 10$ N/kg.

gravitational potential energy = ... J **(2 marks)**

(b) State how much work is done in pulling the log back.

work done = ... J   **(1 mark)**

(c) Calculate the velocity of the battering ram as it reaches the bottom of the swing.

velocity = ... m/s **(4 marks)**

4 Explain the energy changes in a golf ball from when it is hit by the golfer to when it reaches the highest point in the air, in terms of gravitational potential energy and kinetic energy. Identify any energy losses and explain why they may occur.

....

....

....

....

....

.... **(4 marks)**

Extended response – Conservation of energy

Millie plays on a swing in the park. The swing seat is initially pulled back by her friend to 30° to the vertical position and then released. Describe the energy changes in the motion of the swing.

Your answer should also explain, in terms of energy, why the swing eventually stops.

You will be more successful in extended response questions if you plan your answer before you start writing.

The question asks you to give a detailed explanation of the energy changes as the swing moves backwards and forwards. Think about:

- How gravitational potential energy changes as the swing is pulled back.

- Points at where gravitational potential energy (GPE) and kinetic energy (KE) are at maximum and at 0.

- Where some energy may be lost from the system.

- Why the swing will eventually stop.

You should try to use the information given in the question.

....

....

....

....

....

....

....

....

....

....

....

....

.... **(6 marks)**

Waves

1 The table below lists statements about transverse and longitudinal waves. Identify which type of waves are described by writing T (transverse), L (longitudinal) or B (both) next to each statement.

Sound waves are this type of wave.	L	They have amplitude, wavelength and frequency.	x T͟ B
All electromagnetic waves are this type of wave.	T	Seismic S waves are this type of wave.	L
Particles oscillate in the same direction as the wave.	T< +	They transfer energy.	L B +

(3 marks)

2 The diagram below shows a wave travelling through a medium.

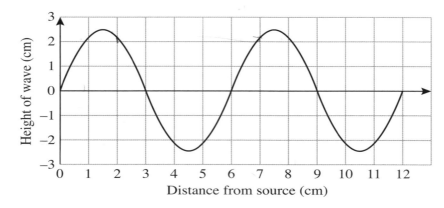

Distance from source (cm)

(a) What is the amplitude of the wave in the diagram?

☐ **A** 0.05 m ☒ **B** 0.025 m ☐ **C** 0.12 m ☐ **D** 0.10 m **(1 mark)**

(b) Determine the wavelength of the wave in the diagram.

> You may find this equation useful $v = f \times \lambda$ or
> wave speed = frequency × wavelength

wavelength = ...5.............. m **(1 mark)**

(c) Sketch a second wave on the diagram to show a higher amplitude and shorter wavelength. **(2 marks)**

3 When a wave travels through a material the average position of the particles of the material remains constant. Explain how this is correct for the types of waves found in:

(a) a sound wave travelling through the air

When a sound wave is generated each particle ....moves back and ferds....

.... **(2 marks)**

(b) ripples travelling across the surface of a pool.

When a water wave is generated the surface ....The particles move the

way the water is going.... **(2 marks)**

Wave equations

Guided

1 Whales communicate over long distances by sending sound waves through the oceans. It takes 20 seconds for the sound waves to travel in seawater between two whales 30 kilometres apart.

Calculate the speed of sound in water in metres/second.

speed = distance travelled by the waves (in metres) ÷ time taken ....

so .... 30.000 ÷ 20 = 150 ....

....

speed of sound = ... m/s **(3 marks)**

2 A sound wave has a wavelength of 0.017 m and a frequency of 20 000 Hz. Calculate the speed of the sound wave in metres/second.

> You may find this equation useful: $v = f \times \lambda$ or
> *wave speed = frequency × wavelength*

20 000 × 0.017 = 340

wave speed = ... m/s **(2 marks)**

3 An icicle is melting into a pool of water. Drops fall every half a second, producing small waves that travel across the water at 0.05 m/s. Calculate the wavelength of the small waves. State the units.

> Remember to write down the equation you are using before you substitute values.

$$\frac{0.05}{0.5} = 0.2s$$

wavelength = ... unit .... **(3 marks)**

4 A satellite sends signals to your TV using radio waves. It takes 0.12 s for the radio waves to travel from the satellite to your TV. The speed of light is 3×10^8 m/s. Calculate the distance of the satellite above the Earth in kilometres.

> You may find this equation useful:
> $v = x \div t$ or *speed = distance ÷ time*

0.12 × 3 × 10⁸ = 36 000

distance = ... km **(3 marks)**

Measuring wave velocity

Guided

1 A tap is dripping into a bath. Three drops fall each second producing small waves that are 5 cm apart. Calculate the speed of the small waves across the water. State the units.

frequency of the waves (f) = ... 7 ....

wavelength of the waves (λ) = ... 5 cm .... 0.05m ....

speed of waves = ... unit .... **(3 marks)**

2 How far apart are the crests of water waves with a frequency of 0.25 Hz travelling at a speed of 2 m/s?

> Your answer to Q1 may help you answer this question.

☐ **A** 0.5 m   ☐ **C** 4 m

☐ **B** 0.125 m   ☑ **D** 8 m   **(1 mark)**

3 A surfer checks out the speed of the waves. He measures the time it takes to surf in a direct line between two buoys 50 m apart. The time taken is 15 seconds. Calculate the speed of the wave. Give your answer to 2 significant figures.

$$S = \frac{d}{t}$$

$$\frac{50}{15}$$

wave speed = ... 3.3 .... m/s **(2 marks)**

4 Eliza and Charlie set up an experiment to estimate the speed of sound in air. They use a brick wall at school and stand 50 metres away. Charlie knocks two pieces of wood together and Eliza measures the time of the echo using a stopwatch.

(a) Describe how Eliza and Charlie can use their measurements to calculate the speed of sound in air.

.... $s = \frac{d}{t}$ .... **(1 mark)**

(b) Suggest three changes that Eliza and Charlie could make to their experiment to make their results more accurate.

.... Repeat it over a range of distances ....

.... use automatic timer ....

....

....

.... **(3 marks)**

Waves and boundaries

1 Use words from the box to complete the sentences.

ray	reflection	refraction
do	do not	normal

Light waves may change direction when they pass from one medium to a different medium.

This is called ....*refraction*... **(1 mark)**

A line drawn at right angles to the boundary of a material is called the........*normal*.... **(1 mark)**

Light waves entering a different medium at 90° ....*do*................... change direction. **(1 mark)**

2 Draw a line to match the behaviour of waves at boundaries with the correct explanation. The first one is done for you.

Wave phenomena	Explanation
reflection	The wave energy is transferred into a thermal energy store.
refraction	The wave bounces back at a surface but does not pass through.
transmission	The wave energy is transferred.
absorption	The wave passes through but at a changed speed.

(2 marks)

3 Explain, in terms of particles, how sound waves travel through air.

> Sound waves are longitudinal waves.

Sound waves are generated by ......*the vibrating ...... that push against air particle*......

.... **(3 marks)**

4 Water waves in a ripple tank can be used to help understand how light waves behave.

(a) Explain what happens when the water waves move from deeper water to shallower water.

......*Water waves travel faster in deep water become ...*......

......*shallow they could also change direction*......

.... **(3 marks)**

(b) Explain what is represented by the change in depth of water when referring to light.

....

....

.... **(1 mark)**

(c) State the cause of the change in direction of both water waves and light waves.

....

.... **(1 mark)**

Waves in fluids

1 A ripple tank is used to investigate waves.

(a) Describe how a ripple tank may be used to measure the frequency of water waves.

....

....

.... **(2 marks)**

(b) Describe how to find the wavelength of the waves in the ripple tank.

....

....

.... **(2 marks)**

(c) State the equation you can use with the data collected in (a) and (b) to determine wave speed.

.... **(1 mark)**

(d) Identify the control variable when using a ripple tank to investigate wave speed.

.... **(1 mark)**

2 Describe a suitable conclusion to the method of using the ripple tank in Q1. Your conclusion should include two factors that should be moderated in this experiment.

Guided

A ripple tank can be used to determine a value for ....

....

.... **(3 marks)**

3 The ripple tank experiment uses several pieces of equipment. Complete the table below to describe the hazard associated with each component and suggest a measure to minimise the risk of harm.

> Identify the hazard and describe the safety measure for each mark.

Component	Hazard	Safety measure
water		
electricity		
strobe lamp		

(3 marks)

Extended response – Waves

A man has dropped his door key into a pond. The key appears to be in a different position because of the phenomenon shown in the diagram. Explain why the key appears to be in a different position.

Your answer should identify the actual position of the key.

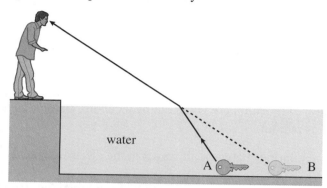

water

A   B

> You will be more successful in extended response questions if you plan your answer before you start writing.
>
> The question asks you to give a detailed explanation of what happens when a light wave passes through the boundary between two different materials. Think about:
>
> - the property of waves that is illustrated
> - the nature of the substances influencing the wave behaviour shown in the diagram
> - the resulting changing direction of the wave as shown in the diagram
> - why the man believes the key is in a different position from the actual position.
>
> You should try to use the information given in the question.

....

....

....

....

....

....

....

....

....

....

....

....

.... **(6 marks)**

Electromagnetic spectrum

1 Microwaves and ultraviolet are types of radiation. Identify the statement that describes these waves correctly.

☐ **A** Microwaves have a higher frequency than ultraviolet.

☑ **B** Microwaves and ultraviolet are transverse waves.

☐ **C** Microwaves have a shorter wavelength than ultraviolet.

☐ **D** Microwaves and ultraviolet are longitudinal waves.   **(1 mark)**

2 Visible and infrared radiation are given out by a candle. Gamma rays are emitted by radioactive elements such as radium.

> **Guided**

   (a) State two similarities between all waves in the electromagnetic spectrum.

All parts of the electromagnetic spectrum are ......transverse...... waves

and they all ...travel the same speed as a vacum...... **(2 marks)**

   (b) Explain why different parts of the electromagnetic spectrum have different properties.

....All different parts carry different energy............... **(1 mark)**

3 The chart below represents the electromagnetic spectrum. Some types of electromagnetic radiation have been labelled.

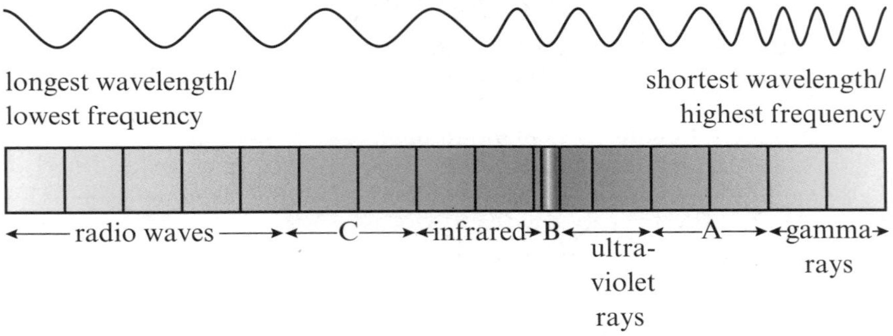

longest wavelength/
lowest frequency

shortest wavelength/
highest frequency

◄——— radio waves ———►◄—C—►◄infrared►B◄———►◄—A—►◄gamma►
                                        ultra-   rays
                                          violet
                                          rays

Name the three parts of the spectrum that have been replaced by letters in the diagram.

A: ...Micro Waves...

B: ...Visible ray...

C: ...x rays...   **(3 marks)**

4 The speed of electromagnetic waves in a vacuum is 300 000 km/s. A radio wave has a wavelength of 240 m. Calculate the frequency of the radio wave.

> 📟 **Maths skills** Remember to convert the units.

$$\frac{300000000}{240} = 1.25 \times 10^8$$

frequency = ...1.25×10^8... Hz   **(3 marks)**

 **Practical skills** # Investigating refraction

1 (a) Suggest a method that could be used to investigate the refraction of light using a glass block and a ray box.

....

....

....

....

....

....

.... **(4 marks)**

Guided (b) Explain what conclusion you would expect to find using the method you have outlined in (a). Your answer should refer to the angle of incidence and the angle of refraction.

When a light ray travels from air into a glass block ....

....

.... **(2 marks)**

(c) (i) Explain what would be observed if the light ray, travelling through the air, entered the glass at an angle of 90° to the surface of the glass.

.... **(1 mark)**

(ii) Explain what would be happening that could not be observed in this experiment.

....

.... **(2 marks)**

2 State three hazards associated with investigating reflection with a ray box and suggest safety measures that could be taken to minimise the risk.

....

....

....

....

.... **(3 marks)**

3 The refraction of light waves through transparent materials can be modelled using a ripple tank.

Describe what changes you would expect to observe in the waves when the depth of the water is made shallower by placing a glass sheet at an angle to the waves.

> Consider the waves being generated in deeper water and then moving into shallower water.

....

....

....

.... **(2 marks)**

Wave behaviour

1 Which of the following is not true about the behaviour of electromagnetic waves?

☐ **A**   Electromagnetic waves travel at 300 000 000 m/s in a vacuum.

☐ **B**   Electromagnetic waves are transverse waves.

☐ **C**   Electromagnetic waves are all transmitted from space through the atmosphere.

☐ **D**   Electromagnetic waves change speed in different materials.   **(1 mark)**

2 (a) Describe four properties of electromagnetic waves that explain their behaviour.

> Recall the properties and nature of waves.

....

....

.... **(4 marks)**

(b) Give examples of two of the properties described in (a).

....

.... **(2 marks)**

3 (a) Explain the differences between microwaves and radio waves that are used for communications.

....

.... **(2 marks)**

> **Guided**

(b) How does the wavelength of these waves dictate the way they are transmitted?

> Long waves are reflected by the ionosphere (part of the atmosphere), while shorter waves pass through.

Microwaves sent from the ground transmitter are ....

....

Radio waves sent from the ground transmitter are ....

.... **(4 marks)**

4 Complete the following statements about the relationship between radio waves and electrical charges.

As .... charges move up and down a (transmitting) radio aerial, oscillating

.... and magnetic fields move from the antenna, across space. When the

oscillating electric .... encounters another (receiving) aerial, it causes oscillations

in the receiving .... circuits.   **(2 marks)**

5 Explain why space-based telescopes have been able to collect extra information about the Universe, beyond that collected by telescopes based on Earth.

....

....

.... **(3 marks)**

Dangers and uses

1   Identify the two correct uses of each type of electromagnetic radiation.

   (a) **infrared**   ☐ **A** night-vision goggles   ☐ **B** broadcasting TV programmes

   ☐ **C** TV remote control   ☐ **D** sun-tan lamps

   > Remember that you need to choose TWO correct answers for each question part (a), (b) and (c).

   (1 mark)

   (b) **ultraviolet**   ☐ **A** thermal imaging   ☐ **B** disinfecting water

   ☐ **C** cooking food   ☐ **D** security marking

   (1 mark)

   (c) **gamma rays**   ☐ **A** sterilising food   ☐ **B** communicating with satellites

   ☐ **C** security systems   ☐ **D** treating cancer

   (1 mark)

2   Complete the following paragraph by circling the correct word in each case.

   Some electromagnetic waves can be dangerous. **Microwaves/Ultraviolet** can **heat/freeze** the water inside our bodies causing significant damage to cells. **Infrared/Visible light** waves transfer **thermal/chemical** energy and can cause burns to skin. **Ultraviolet/Radio** waves can damage **eyes/ears** and can cause skin cancer.

   (3 marks)

3   Describe how living cells might be affected following over-exposure to X-rays or gamma rays and what might occur at a cellular level as a result of this.

   > X-rays and gamma waves move as high-energy ionising photons.

   ....

   ....

   ....

   .... **(2 marks)**

4   X-ray scans are often taken when athletes are injured, but X-rays are known to be harmful. Discuss the use of X-rays in the treatment of injuries.

   Guided

   X-rays are useful because ....

   X-rays can be harmful ....

   The use of X-rays should be controlled by ....

   ....

   ....

   .... **(3 marks)**

Changes and radiation

1 Which of the following statements about electrons is true?

☐ **A** Electrons only change orbit when they emit electromagnetic radiation.

☐ **B** Electrons only absorb electromagnetic radiation.

☐ **C** Electrons always change orbit when electromagnetic radiation is absorbed.

☐ **D** Electrons can only orbit the nucleus at defined energy levels within the atom.   **(1 mark)**

2 Electromagnetic radiation can be absorbed by electrons that orbit an atomic nucleus.

(a) Explain why different types of electromagnetic radiation have different energies.

....

.... **(2 marks)**

> **Guided**

(b) Describe what happens to an electron that absorbs electromagnetic radiation.

When an electron absorbs electromagnetic radiation ....

.... **(1 mark)**

> **Guided**

(c) Describe what happens to an electron that emits electromagnetic radiation.

When an electron emits electromagnetic radiation ....

.... **(1 mark)**

3 The diagram shows different electron energy levels in an atom. Describe what an 'excited' electron is and why it moves to a lower energy level, emitting a photon of a certain wavelength only. Use the diagram to help you.

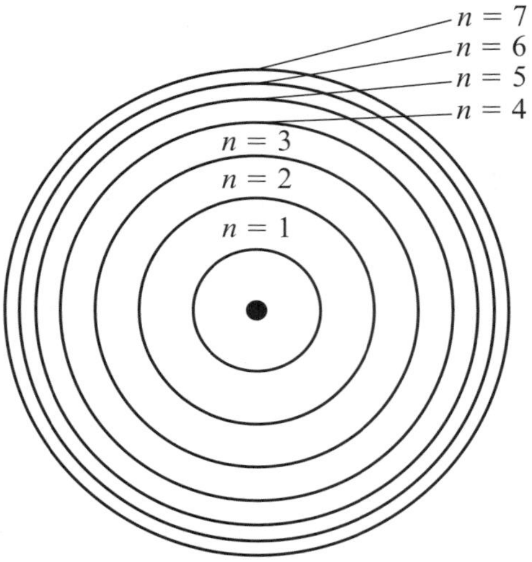

> Electrons can only orbit the nucleus of an atom at certain energy levels.
>
> Electrons gain a specific amount of energy from photons to move to a higher energy level.

$n = 7$
$n = 6$
$n = 5$
$n = 4$
$n = 3$
$n = 2$
$n = 1$

....

....

....

.... **(3 marks)**

4 Explain why photons can also be emitted from the nucleus of an unstable atom and why these have much higher frequencies than those emitted from orbiting electrons.

....

....

.... **(3 marks)**

Extended response – Light and the electromagnetic spectrum

Electromagnetic waves are used in industry in a wide variety of applications.

Describe the harmful effects that can result in over-exposure to certain types of waves.

Suggest ways of reducing damage to body cells from these waves.

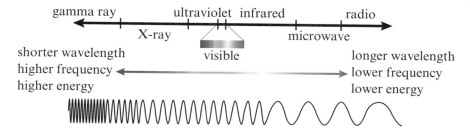

> You will be more successful in extended response questions if you plan your answer before you start writing.
>
> The question asks you to give a detailed explanation of the harmful effects that may result from excessive exposure to electromagnetic radiation. In your answer, explain:
>
> - the heating effect of microwaves
> - the dangers from infrared radiation
> - initial and long-term damage from ultraviolet waves
> - effects on body cells from the very high energy waves, X-rays and gamma waves
> - how the dangers of exposure to ultraviolet and infrared waves could be reduced.
>
> You should try to use the information given in the question and in the diagram.

....

....

....

....

....

....

....

....

....

....

....

....

.... **(6 marks)**

Structure of the atom

1 Complete the diagram to show the location and charge of:

(a) protons *nucleus* **(1 mark)**

(b) neutrons *nucleus* **(1 mark)**

(c) electrons. *the outer shell* **(1 mark)**

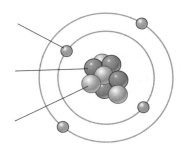

2 (a) Explain why atoms have no overall charge.

...... *nucleus + neutrons + protons + electron* ......

.... **(2 marks)**

(b) State what will happen to the overall charge if an atom loses an electron.

.... *+1* .... **(1 mark)**

3 (a) State what is meant by the term molecule.

.... **(1 mark)**

(b) Give an example of:

(i) a molecule of liquid

> Remember that molecules are given a chemical formula.

.... **(1 mark)**

(ii) a molecule of a gaseous element

.... **(1 mark)**

(iii) a molecule of a gaseous compound.

.... **(1 mark)**

4 The diagram shows an atomic nucleus, an atom and a molecule. Choose the closest approximate size for each and write it under the diagram.

10^{-10} m 10^{-18} m 10^{-9} m 10^{-15} m 10^{-2} m 10^{-6} m

nucleus   atom   molecule

....

(3 marks)

Atoms and isotopes

Guided

1 State what is meant by each term:

(a) nucleon: *the name given to particles in the* .... **(1 mark)**

(b) atomic number: .... **(1 mark)**

(c) mass number: .... **(1 mark)**

2 Identify the correct description of isotopes.

☐ **A** atoms of the same element with different numbers of electrons

☐ **B** atoms of different elements with same numbers of neutrons

☐ **C** atoms of the same element with different numbers of neutrons

☐ **D** atoms of the same element with different numbers of protons **(1 mark)**

3 Explain why different isotopes of the same element will still be electrically neutral.

> Consider all the particles of the isotopes.

....

.... **(2 marks)**

4 State and explain the full symbol of a potassium atom (K) that has 19 protons and 20 neutrons in its nucleus.

....

.... **(2 marks)**

5 Identify the correct statement about the relative mass of particles in an atom.

☐ **A** proton relative mass 1, neutron relative mass 0, electron relative mass −1

☐ **B** proton relative mass +1, neutron relative mass 0, electron relative mass negligible

☐ **C** proton relative mass 1, neutron relative mass 1, electron relative mass negligible

☐ **D** proton relative mass 1, neutron relative mass 1, electron relative mass −1 **(1 mark)**

6 The full symbols of three isotopes of lithium are shown below.

$^{6}_{3}\text{Li}$ $^{7}_{3}\text{Li}$ $^{8}_{3}\text{Li}$

Compare the structures of the three isotopes in terms of protons, neutrons and nucleons.

> Consider the difference between the mass number and the atomic number.

....

....

....

.... **(3 marks)**

Atoms, electrons and ions

1 Select the correct statement about atoms.

    ☐ **A** Electrons orbit at fixed distances from the nucleus.

    ☐ **B** Electrons orbit the nucleus at random distances from the nucleus.

    ☐ **C** An electron can be lost from the atom when it emits electromagnetic radiation.

    ☐ **D** Electrons move to a lower orbit when they absorb electromagnetic radiation.

(1 mark)

2 Describe what happens to an electron when its atom:

  (a) absorbs electromagnetic radiation

> **Guided**

When an atom absorbs electromagnetic radiation ....

.... **(2 marks)**

  (b) emits electromagnetic radiation.

When an atom emits electromagnetic radiation ....

.... **(2 marks)**

3 (a) Write these in the correct part of the table.

    F^- Li Na^+ B^+ K^+ Cu

Atoms	Ions

(2 marks)

  (b) Explain your choices. | Describe the influence of electrons on both atoms and ions. |

    ....

    ....

    ....

    .... **(2 marks)**

4 Explain two ways in which a neutral atom can become a positive ion through losing an electron. You may sketch a diagram to help you explain your answer.

| Consider the effect of ionising radiation. |

    ....

    ....

    ....

    .... **(2 marks)**

Ionising radiation

1   Select the correct description of an alpha particle.

    ☐   **A**   helium nucleus with charge −2

    ☐   **B**   helium nucleus with charge +2

    ☐   **C**   high-energy neutron

    ☐   **D**   ionising electron

(1 mark)

2   Match the types of radiation with the correct penetrating power.

Type of radiation
alpha
beta minus
neutron
gamma

Penetrating power
low, stopped by thin aluminium
high
very high, stopped by very thick lead
very low, stopped by 10 cm of air

(3 marks)

3   An atom of carbon-14, with 6 protons and 8 neutrons, undergoes beta-minus decay to become an atom of nitrogen, with 7 protons and 7 neutrons.

(a)   Give the change in relative atomic mass.

.... **(1 mark)**

(b)   Give another description for a beta-minus particle.

.... **(1 mark)**

(c)   State the relative ionising property of a beta-minus particle.

.... **(1 mark)**

4   Identify the type of radiation that would be emitted in each decay.

> Remember the law of conservation of mass.

(a)   carbon-10 (6 protons, 4 neutrons) ➔ boron-10 (5 protons, 5 neutrons)

.... **(1 mark)**

(b)   uranium-238 (92 protons, 146 neutrons) ➔ thorium-234 (90 protons, 144 neutrons)

.... **(1 mark)**

(c)   helium-5 (2 protons, 3 neutrons) ➔ helium-4 (2 protons, 2 neutrons)

.... **(1 mark)**

5   Explain why alpha particles have the shortest ionising range in air, compared to other types of ionising radiation.

Guided

Compared to other types of ionising radiation, the chance of collision with air particles

....

because ....

.... **(3 marks)**

Background radiation

1 The pie chart shows the sources of background radiation. 50% of this comes from the element radon. Explain what radon is and how it occurs.

....

.... **(2 marks)**

ground and buildings
14.0%

medical
14.0%

nuclear power
0.3%

cosmic rays
(from space)
10.0%

other
0.2%

food
and drink
11.5%

radon
gas
50.0%

2 Give two reasons why radon levels can vary across the UK.

Guided

Levels can vary because of the different rocks ....

They can also vary .... **(2 marks)**

3 Complete the table by giving examples of natural and man-made sources of background radiation.

Sources of background radiation	
Natural	**Man-made**

(2 marks)

4 A scientist in the south-east of England measures the background radiation count three times. Her colleague conducts the same experiment in the south-west. Their results are shown in the table below.

Test number	1	2	3	Average
south-east activity (Bq)	0.30	0.24	0.27	
south-west activity (Bq)	0.31	0.28	0.32	

(a) Calculate the average activity for each sample and write it in the table. **(2 marks)**

(b) State which area has the highest level of background radiation. **(1 mark)**

....

5 (a) Describe how radon gas can get into homes and buildings.

....

.... **(2 marks)**

(b) Explain why inhaling radon can be more dangerous than external exposure.

Consider the type of
radiation produced.

....

.... **(2 marks)**

Measuring radioactivity

Guided

1 In 1896 Henri Becquerel discovered that uranium salts led to a darkening of photographic film. Explain how this is used in the nuclear industry today.

Photographic film is used by nuclear industry workers ....

....

This monitors levels of .... **(3 marks)**

Guided

2 Complete the flow chart below to show how a Geiger–Muller tube detects nuclear radiation.

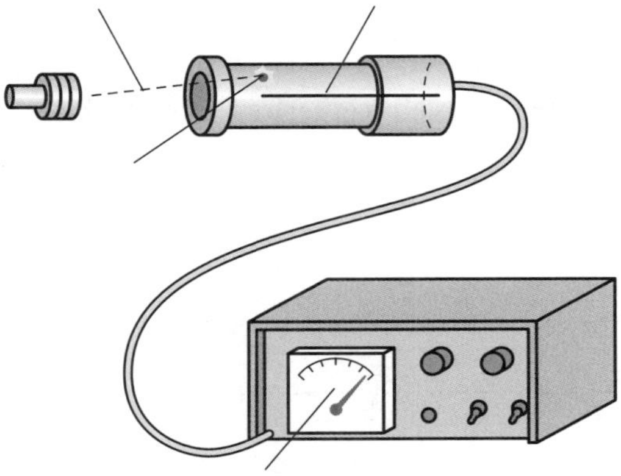

A thin wire	→	Atoms of argon are	→	Electrons travel	→	The amount of

(4 marks)

3 A student claims that the more ionising the radiation, the more effective the G–M tube is at detecting levels of radiation. Explain whether the student is correct.

....

....

....

.... **(3 marks)**

4 Explain why photographic film badges have aluminium and lead sheets inserted in front of different parts of the photographic film.

> Aluminium will absorb some beta particles and lead will absorb some gamma waves. Absorption will depend on the thickness of the materials.

....

....

.... **(3 marks)**

Models of the atom

1 Compare and contrast the plum pudding and Rutherford models of the atom.

> **Guided**

The plum pudding model showed the atom as ....

....

....

while the Rutherford model showed the atom as ....

....

.... **(4 marks)**

2 Describe the evidence that enabled Rutherford to make his claim about the nucleus.

....

....

....

.... **(3 marks)**

3 (a) What particle did Niels Bohr's model of the atom specifically develop new ideas for?

☐ **A** the electron   ☐ **C** the neutron

☐ **B** the proton   ☐ **D** the nucleus   **(1 mark)**

(b) Explain how the Bohr model of the atom improved on Rutherford's model.

....

....

....

....

....

.... **(4 marks)**

4 (a) Identify the cause of orbiting electrons moving to a higher energy level within the atom.

> Electrons orbit only at certain energy levels. They can gain specific amounts of energy from EM radiation.

....

....

.... **(2 marks)**

(b) When an electron loses energy, it can fall to a lower energy level. State the other outcome when an electron loses energy.

....

.... **(1 mark)**

Beta decay

1 Draw lines to link the boxes to complete the sentences about beta-minus and beta-plus decay. One has been done for you.

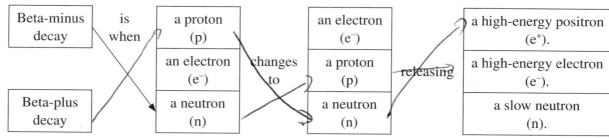

Beta-minus decay	is when	a proton (p)		an electron (e⁻)		a high-energy positron (e⁺).
Beta-plus decay		an electron (e⁻)	changes to	a proton (p)	releasing	a high-energy electron (e⁻).
		a neutron (n)		a neutron (n)		a slow neutron (n).

(2 marks)

2 Using the data in the table, complete the equations below.

7 Li lithium 3	9 Be beryllium 4	11 B boron 5	12 C carbon 6	14 N nitrogen 7	16 O oxygen 8	19 F fluorine 9	20 Ne neon 10
23 Na sodium 11	24 Mg magnesium 12	27 Al aluminium 13	28 Si silicon 14	29 P phosphorus 15	31 S sulfur 15	35.5 Cl chlorine 17	40 Ar argon 18

(a) $^{14}_{6}C \rightarrow {}^{14}_{7}N + {}^{0}_{-1}e$ **(1 mark)**

(b) $^{23}_{12}Mg \rightarrow {}^{23}_{11}Na + {}^{0}_{1}e$ **(1 mark)**

3 Describe what happens in:

> In beta decay neutrons and protons undergo changes producing high energy beta particles. Describe these changes and the particles emitted.

(a) beta-minus decay

..... is when a neutron changes to an electron and proton the proton stays in the nucleus while the **(2 marks)** electron becomes high energy

(b) beta-plus decay.

.. is when a proton changes to a neutron releasing a high energy positron ..... **(2 marks)**

4 Elements that undergo beta decay are used in archaeological and medical applications. Describe how these professions use beta decay.

Guided

In archaeology, beta decay is used to ....

In medicine, beta decay is used for .. PET   Scans .... **(2 marks)**

Radioactive decay

1   Radium-222 undergoes alpha decay. Identify which **two** of the following statements are true.

☐   **A**   The positive charge of the nucleus is reduced by 4.

☐   **B**   The mass number is reduced by 4.

☐   **C**   The atomic number is reduced by 2.

☐   **D**   The nucleus gains an extra proton.   **(1 mark)**

2   Beta decay has two forms. Name the two types of beta decay and give the charge for each type.

Guided

beta- ...., charge ....

beta- ...., charge ....   **(2 marks)**

3   Explain how conserving mass number relates to nuclear decay.

....

....

....   **(2 marks)**

4   Describe what happens in neutron decay.

....

....

....   **(2 marks)**

5   Identify the correct term that could be used in a description of gamma decay.

☐   **A**   electron   ☐   **B**   photon   ☐   **C**   positron   ☐   **D**   proton   **(1 mark)**

6   (a)   Complete each equation and state what type of decay is shown.

> Check that the A and Z numbers obey the conservation laws.

(i) $^{208}_{84}\text{Po} \rightarrow ^{4}_{2}\text{He} + ^{204}_{82}\text{Pb}$ type of decay ...... *Alpha* ......   **(2 marks)**

(ii) $^{222}_{}\text{Rn} \rightarrow ^{4}_{2}\text{He} + ^{218}_{84}\text{Po}$ type of decay ...... *Alpha* ......   **(2 marks)**

(iii) $^{42}_{19}\text{K} \rightarrow ^{0}_{-1}\text{e} + ^{}_{20}\text{Ca}$ type of decay ...... *Beta⁻* ......   **(2 marks)**

(iv) $^{}_{4}\text{Be} \rightarrow ^{1}_{0}\text{n} + ^{8}_{4}\text{Be}$ type of decay ...... *Beta⁺* ......   **(2 marks)**

(b)   Nuclear decay results in a loss of energy from the nucleus. State the reason for this and discuss the energy transfer involved.

....

....   **(2 marks)**

Half-life

1 A sample of thallium-208 contains 16 million atoms. Thallium-208 has a half-life of 3.1 minutes.

> Half-life is the time for half the nuclei in a sample to decay, not the time for one atom to decay.

(a) State the number of nuclei that will have decayed in 3.1 minutes.

number of nuclei = ... **(1 mark)**

(b) Calculate the number of unstable thallium nuclei left after 9.3 minutes.

number of unstable thallium nuclei left = ... **(2 marks)**

Guided

2 A student measured the activity of a radioactive sample for 30 minutes. She plotted the graph of activity against time shown below.

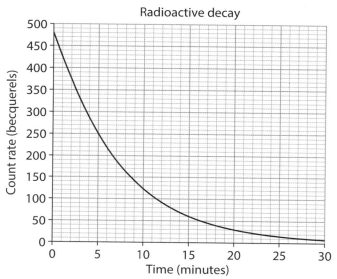

Use the graph to calculate the half-life of the sample.

> You could take any point on the line as a starting point for calculating half-life.

The activity is .... Bq at .... min.

Half this activity is .... Bq, which is at .... min

so the half-life is ....

half-life = ... min **(3 marks)**

3 After the Chernobyl nuclear power station exploded in 1986 a radioactive isotope, caesium-137, fell on northern England and Wales. At one place the activity of a soil sample was 64 Bq in 1986. The radioactivity due to caesium-137 was expected to fall to 32 Bq in 30 years. The level of radioactivity in the soil is higher than predicted. Discuss the factors that may have influenced the accuracy of this prediction.

....

....

.... **(3 marks)**

Dangers of radiation

1   The hazard symbol shown is used to warn that sources of ionising radiation may be present.

Give two places where this symbol may be displayed.

1   ....

2   .... **(2 marks)**

2   (a) State what is meant by the term ionising.

   Ionising means .... **(1 mark)**

   (b) Explain why ions are dangerous in the body.

   Ions in the body can ....

   which can lead to .... **(2 marks)**

3   (a) Describe how employers using radioactive sources can take steps to reduce the exposure of their workers to ionising radiation.

   ....

   ....

   .... **(3 marks)**

   (b) When risk of exposure to radiation has been minimised through procedure, describe how workers can be monitored to further improve their safety.

> Photographic film is an important tool in monitoring levels of exposure to radiation. Think how this is used.

   ....

   ....

   .... **(2 marks)**

4   Radioactive sources that are used in schools are handled following special procedures.

   Give a reason for the use of the following:

   (a) long-handled tongs to move radioactive sources

   ....

   .... **(1 mark)**

   (b) lead-lined boxes for storing radioactive sources

   ....

   .... **(1 mark)**

5   Explain why X-rays are classed as ionising.

   ....

   .... **(2 marks)**

Contamination and irradiation

1   During the First World War (1914–18) soldiers and airmen were issued with watches that had hands and numbers that glowed in the dark. The hands and numbers had been painted with luminous paint that contained radium. Radium was discovered in 1898 and found to be radioactive. In the 1920s many of the women who painted the watches became very ill.

(a)   Explain why radium paint was used to paint the watches.

.... **(1 mark)**

Guided

(b)   Discuss why it was not banned from being used on watches until the 1920s.

Before 1920 the effects of radium ....

....

....

.... **(3 marks)**

2   Draw a line from each term to its correct description.

Term	Description
external contamination	A radioactive source is eaten, drunk or inhaled.
internal contamination	A person becomes exposed to an external source of ionising radiation.
irradiation	Radioactive particles come into contact with skin, hair or clothing.

(2 marks)

3   Give an example of how a person may be subjected to:

(a)   external contamination

.... **(1 mark)**

(b)   internal contamination.

.... **(1 mark)**

4   Explain why alpha particles are more dangerous from a source of internal contamination than from a source of irradiation.

> Alpha particles can only travel short distances before they collide with another particle and lose their energy. This can have serious consequences near to the body.

....

....

....

....

....

.... **(4 marks)**

Extended response – Radioactivity

Ionising radiation will travel through some materials but will be stopped by other materials. The diagram shows three different materials and how alpha, beta or gamma radiation may be stopped.

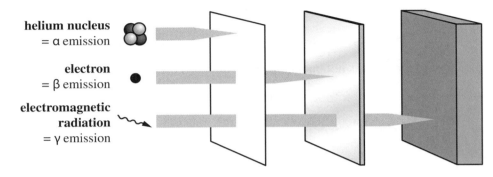

helium nucleus = α emission

electron = β emission

electromagnetic radiation = γ emission

Explain what the diagram shows about the nature of these three types of radiation. In your answer, name the materials that could be shown in the diagram, and give other examples of how these radiations may be stopped.

You will be more successful in extended response questions if you plan your answer before you start writing.

The question asks you to give a detailed explanation of the penetrative characteristics of ionising radiation and examples of materials that may stop the radiation. Think about:

- the relative ionising abilities of alpha, beta and gamma radiation
- how energy is transferred when radiation encounters a particle
- the effect that particle collisions have on how radiation passes through a material
- examples of materials that stop different types of radiation.

You should try to use the information given in the diagram.

....

....

....

....

....

....

....

....

....

....

....

.... **(6 marks)**

Continue your answer on your own paper. You should aim to write about half a side of A4.

Work, energy and power

1 A kettle has a power rating of 2500 W. How much energy does it use in 5 seconds?

☐ **A** 2500 J

☐ **B** 500 J

☐ **C** 500 W

☐ **D** 12 500 J   **(1 mark)**

2 Give the energy store that increases in each of these examples:

> Energy transfers result in the movement of energy from one store to another.

(a) when a mass is lifted through a height

.... **(1 mark)**

(b) when a pan of water at 20 °C water is heated to 70 °C

.... **(1 mark)**

(c) when an extra cell is added to a circuit.

.... **(1 mark)**

Guided

3 A microwave cooker heats a drink in 20 seconds using 15 000 J of electrical energy. Calculate the power of the microwave cooker. State the unit.

energy transferred = ... J, time taken = ... s

$P = \dfrac{E}{t} =$

power = ... unit .... **(3 marks)**

4 A student weighing 600 N climbs 20 stairs to a physics lab. Each stair is 0.08 m high. Calculate the work done by her muscles to climb the stairs. State the unit.

work done = ... unit .... **(3 marks)**

5 A student watches a programme on his television, which has a power rating of 200 W and uses 360 000 J of energy during the viewing. Calculate the time the student spends watching the television.

time taken = ... s **(3 marks)**

Extended response – Energy and forces

Wind turbines are designed to use the kinetic energy of moving air to turn the turbine blades. Explain why this is described as a mechanical process and, therefore, why efficiency is important.

> You will be more successful in extended response questions if you plan your answer before you start writing.
>
> The question asks you to give a detailed explanation of the mechanical processes involved in energy transfers in a wind turbine. Think about:
>
> - the meaning of 'mechanical process'
> - the parts of the process where mechanical processes occur
> - how mechanical processes are inefficient and examples of this
> - methods of reducing wasted energy due to mechanical processes.
>
> You should try to use the information given in the question.

....

....

....

....

....

....

....

....

....

....

....

....

....

.... **(6 marks)**

Interacting forces

1 (a) Give the three types of fields that cause objects to interact with each other **without** making contact.

…………………………..   ……………………..……   ………….…………….….   **(3 marks)**

(b) Explain which of these is different from the other two and why.

> Two of these fields have opposite poles or charges but one acts in only one direction.

....

....

....

.... **(2 marks)**

2 Which **two** of the following correctly describe the similarities between magnetic and electrostatic fields?

☐ **A** Like poles/charges repel.

☐ **B** Like poles/charges attract.

☐ **C** Opposite poles/charges attract.

☐ **D** Opposite poles/charges repel **(2 marks)**

3 Explain why weight and normal contact force are described as vectors.

Weight is a vector because ....

....

Normal contact force is a vector because ....

.... **(2 marks)**

4 A student pulls along a luggage bag, as shown in the diagram, at constant velocity.

(a) Identify the contact forces for the horizontal motion and state which force is larger.

....

.... **(3 marks)**

(b) Name the balanced contact forces in the vertical direction.

.... **(1 mark)**

Free-body force diagrams

1 Which of the following is **not** one of the steps for resolving a resultant force using a scale drawing?

☐ **A** The angles are drawn with a protractor.

☐ **B** The scale is not important.

☐ **C** The vertical component can be found through measurement.

☐ **D** The horizontal line is drawn to scale. **(1 mark)**

2 The diagram shows a bird of prey of mass 2 kg on a branch of a tree.

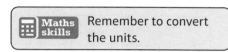

Maths skills — Remember to convert the units.

(a) Add arrows to the diagram to show the direction of forces acting on the bird. **(2 marks)**

(b) Add the magnitude and units for each force to the diagram. **(4 marks)**

3 Draw a free-body diagram to represent a cyclist accelerating to the left along a horizontal path. **(4 marks)**

Remember to consider the relative length of the arrows drawn.

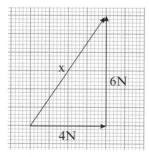

4 The diagram shows the horizontal and vertical components of a force.

x

6N

4N

(a) Measure the length of the line of the resultant force x. .... cm **(1 mark)**

(b) Give the magnitude of the resultant force x. .... N **(1 mark)**

Resultant forces

1   Below are diagrams of pairs of forces.

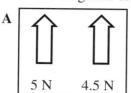 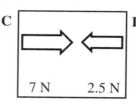 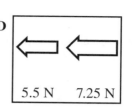

resultant = ... N   resultant = ... N   resultant = ... N   resultant = ... N

direction ....   direction ....   direction ....   direction ....

(a)   Calculate the value of the resultant force for each pair.   **(4 marks)**

(b)   Add an arrow to show the direction of each resultant force.   **(4 marks)**

2   The diagram shows two force pairs.

What is the magnitude of the resultant force?

☐ **A** 3 N

☐ **B** 4 N

☐ **C** 5 N

☐ **D** 7 N

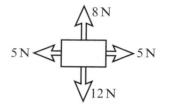

(1 mark)

3   The diagram represents two component forces of a resultant force.

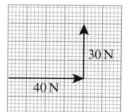

(a)   Add an arrow to represent the resultant force.

(1 mark)

(b)   Calculate the force represented by the resultant line.   .... N   **(1 mark)**

4   Determine the resultant force acting on a hockey ball with component forces of 15 N acting horizontally and 6 N acting vertically. Draw a scale diagram to calculate your answer.

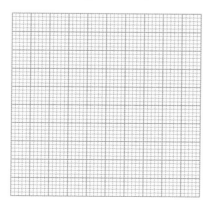

Check your scale before you start to draw the diagram.

(3 marks)

217

Extended response – Forces and their effects

Balanced and resultant forces are an important consideration when operating a camera drone remotely. To film an aerial sequence, the camera operator flies the drone to the required height for filming and keeps it stationary by controlling the speed of the rotor blades. Explain the balanced or unbalanced forces acting on the camera drone during operation (assume breeze is negligible).

> You do **not** need to explain thrust or aerodynamics.

> You will be more successful in extended response questions if you plan your answer before you start writing.
>
> The question asks you to give a detailed explanation of the forces acting on the drone as it is flown to the position for filming. Think about:
>
> - the lift force from the rotor blades
> - how resultant forces lead to motion of an object
> - the vertical and horizontal forces and how they are balanced or unbalanced
> - the forces that the drone must overcome to be able to fly or hover.
>
> You should try to use the information given in the question.

....

....

....

....

....

....

....

....

....

....

....

....

.... **(6 marks)**

Circuit symbols

1 State the reason why a resistor heats up as a result of an electric current flowing through it.

.... **(1 mark)**

2 (a) Select the **two** components that respond automatically to changes in the environment.

☐ **A** diode   ☐ **C** LDR

☐ **B** LED   ☐ **D** thermistor   **(2 marks)**

Guided

(b) Describe how the components you have chosen respond to changes in the environment.

(i) The .... responds by .... **(1 mark)**

....

(ii) The .... responds by .... **(1 mark)**

....

3 Complete the table of circuit symbols below.

Component	Symbol	Purpose
ammeter		
		provides a fixed resistance to the flow of current
	 ─▷⊢─	
		allows the current to be switched on or off

(4 marks)

4 Devise a circuit that could be used to operate a motorised fan to start when the temperature gets too hot. The operator should be able to take readings of the motor voltage and the circuit current. There should also be an option to turn the fan off manually.

Consider whether each component should be connected in series or in parallel.

(6 marks)

Series and parallel circuits

1 (a) Each lamp in these circuits is identical. Write the current for each ammeter on the circuit diagrams.

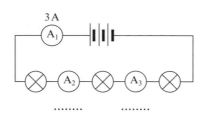

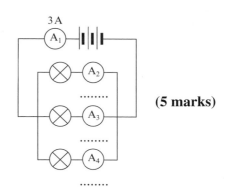

(5 marks)

Guided

(b) Explain the rules for current in series and parallel circuits.

In a series circuit the current ....

In a parallel circuit the current .... **(2 marks)**

2 (a) Each lamp in these circuits is identical. Write the potential difference for each voltmeter in the circuit diagrams.   **(6 marks)**

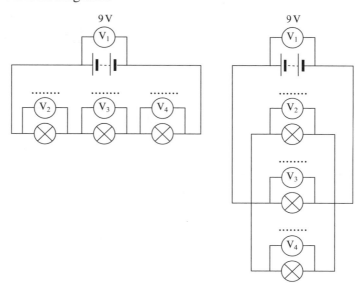

Guided

(b) Explain the rules for potential difference in series and parallel circuits.

In a series circuit the potential difference ....

In a parallel circuit the potential difference .... **(2 marks)**

3 (a) Explain why the electricity supply in a building is connected as a parallel circuit.

> Buildings supply electricity for a number of different appliances that are often used at the same time. The supply must be able to support these as well as providing a reliable source of electricity.

....

.... **(2 marks)**

(b) Give two consequences of connecting lamps in a series circuit.

....

.... **(2 marks)**

Current and charge

1 The electric current flowing in a circuit is 4 A.

(a) Explain what is meant by an electric current.

....

.... **(2 marks)**

(b) The current flows for 8 seconds. Calculate how much charge has flowed. Give the unit.

> You may find this equation useful:
> $Q = I \times t$

charge = ... unit .... **(3 marks)**

2 The diagram shows a series circuit.

(a) Give the reading on ammeter:

(i) A_1 .... A

(ii) A_3 .... A

0.3 A

(1 mark)

(1 mark)

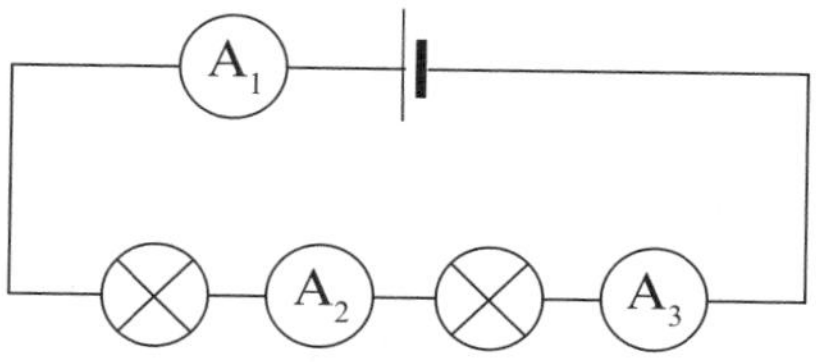

(b) State how you could increase the size of the current flowing through the circuit.

.... **(1 mark)**

> **Guided**

(c) Explain why the current measured by ammeter A_2 is the same as A_1 and A_3.

The electrons move around the ....

so the current leaving the cell is the same as ....

returning to it.

(2 marks)

3 A student is investigating how current carries charge around a circuit.

(a) Draw a circuit diagram to show how the charge could be measured.

(2 marks)

(b) State what else the student would need to use to collect enough data in order to calculate charge.

.... **(1 mark)**

Energy and charge

Guided

1 State what is meant by current and potential difference. Include the word 'charge' in your answers.

Current is the ....

Potential difference is the .... **(2 marks)**

Guided

2 Calculate the amount of energy transferred to a 9 V lamp when a charge of 30 C is supplied.

charge = ... C

potential difference = ... V

so E = ...

> You may find this equation useful:
>
> $E = Q \times V$

energy transferred = ... J   **(3 marks)**

3 Calculate the charge needed to transfer 125 J of energy to a string of fairy lights with a total potential difference of 5 V.

> In questions like these, you will need to rearrange familiar equations.
>
> In this case, $E = Q \times V$ will become $Q = E \div V$

charge = ... C   **(3 marks)**

4 Calculate how long it takes for 600 J of electrical energy, carried by a current of 0.15 A, to be transferred by a resistor with a potential difference of 20 V.

> **Maths skills** This question will need a two-step answer. Step one, calculate charge using $Q = E \div V$.
> Step two, calculate the time taken using $Q = I \times t$.
> You will need to be able to rearrange this second equation to give $t = Q \div I$.

time = ... s   **(4 marks)**

Ohm's law

1 Which quantity is the ohm (Ω) a unit of?

☐ **A** current   ☐ **C** potential difference

☐ **B** energy   ☐ **D** resistance   **(1 mark)**

2 Explain what Ohm's law means.

> Guided

Ohm's law means that ....

is directly proportional .... **(2 marks)**

3 Calculate the resistance of each resistor.   You may find this equation useful: $R = V \div I$

(a) A resistor with a potential difference of 12 V across it and a current of 0.20 A passing through it.

resistance = ... Ω   **(2 marks)**

(b) A resistor with a potential difference of 22 V across it and a current of 0.40 A passing through it.

resistance = ... Ω   **(2 marks)**

(c) A resistor with a potential difference of 9 V across it and a current of 0.03 A passing through it.

resistance = ... Ω   **(2 marks)**

(d) Identify which resistor has the highest resistance.

.... **(1 mark)**

4

(a) Sketch two lines on the graph to show two ohmic conductors of different resistances labelled A and B.   **(2 marks)**

(b) From your graph, identify which line represents the resistor with the higher resistance.

....

.... **(1 mark)**

Resistors

1 The diagrams show resistors in series and resistors in parallel. The resistors in both circuits are identical.

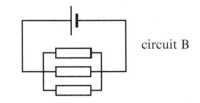

Series circuits: $R_T = R_1 + R_2 + R_3$
Parallel circuits: $1/R_T = 1/R_1 + 1/R_2 + 1/R_3$

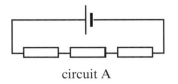

circuit A

circuit B

Identify the correct description of the total resistance in circuit B.

☐ **A** It is lower than in circuit A.

☐ **B** It is higher than in circuit A.

☐ **C** It is the same as in circuit A.

☐ **D** The resistance is variable. **(1 mark)**

2 The diagram below shows a 20 Ω resistor, a 30 Ω resistor and a 150 Ω resistor connected in series with identical cells. The current measured by the ammeter is 0.03 A.

(a) Calculate the total resistance of the circuit.

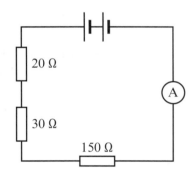

total resistance = ... Ω **(2 marks)**

> **Guided**

(b) (i) State the rule for potential difference in this type of circuit.

.... **(1 mark)**

(ii) Calculate the potential difference across each cell.

Step 1: Calculate total potential difference of the circuit using V = I × R

where I = ... A   and R = (your answer from (a)) .... Ω

So V = ... = ... V

Step 2: Divide the answer at Step 1 by the number of cells in the series circuit.

potential difference = ... V **(3 marks)**

3 A circuit is supplied with 0.9 V. The current flowing through the 15 Ω resistor is 0.06 A. The current flowing through the 10 Ω resistor is 0.09 A. Explain why these values for current are different by giving a calculated example.

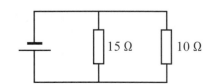

....

....

.... **(4 marks)**

I–V graphs

1   The graphs below show three types of component.

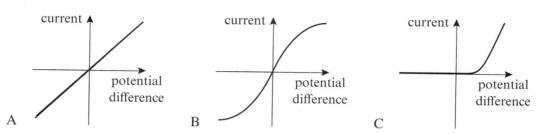

  (a)   Which graph shows the characteristics of a diode? .... **(1 mark)**

  (b)   Describe what happens to the current through the component shown in graph A as the potential difference increases.

     ....

     .... **(2 marks)**

  (c)   Describe what happens to the current through the component shown in graph B as the potential difference increases.

     ....

     .... **(2 marks)**

2 (a)   Complete the I–V graphs for a fixed resistor (at constant temperature) and a filament lamp.

                                                                                     (2 marks)

              Fixed resistor   Filament lamp

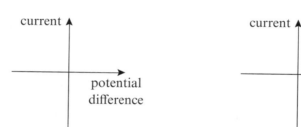

  (b)   Explain why the filament lamp graph has a different shape to the fixed resistor graph.

                                     > A fixed resistor (at constant temperature) obeys Ohm's Law but a filament lamp does not.

     ....

     .... **(2 marks)**

3   Describe an experiment to collect data to enable the calculation of resistance. You may sketch a diagram to illustrate your answer.

     Data can be collected using an ....

     and a ....

     A .... should be included which will allow

     ....

     Resistance can then be calculated from .... **(5 marks)**

 **Practical skills**

Electrical circuits

Guided

1   Electrical circuits can be connected in series or in parallel. Draw a circuit diagram below to show how you could measure the current and potential difference of two resistors connected in a:

   (a)   series circuit   **(3 marks)**

> Remember to include the ammeters and voltmeters in your diagrams.

   (b)   parallel circuit.   **(3 marks)**

   (c)   Describe the differences in the current and potential difference in a series and a parallel circuit.

      ....

      ....

      ....

      ....

      .... **(4 marks)**

2   (a)   Describe how a circuit may be set up to investigate the relationship between current and potential difference using a filament lamp.

      ....

      ....

      .... **(3 marks)**

   (b)   State the law that may be applied to the data collected in (a) to find resistance.

      .... **(1 mark)**

   (c)   Describe how resistance may be found by drawing a graph using the data that would be obtained from the circuit in (a).

      ....

      .... **(2 marks)**

3   Identify a safety hazard when using resistors.

      .... **(1 mark)**

The LDR and the thermistor

1   Draw the circuit symbols for the components in the boxes provided.

Light-dependent resistor (LDR)	Thermistor

(2 marks)

2   Which variable do you need to change to get a change in the resistance of a thermistor?

☐   **A**   current

☐   **B**   humidity

☐   **C**   light

☐   **D**   temperature   **(1 mark)**

3   Each sketch graph below shows the relationship between two variables.

> **Practical skills**   Recall independent and dependent variables.

     (a)   Describe how resistance changes with light.

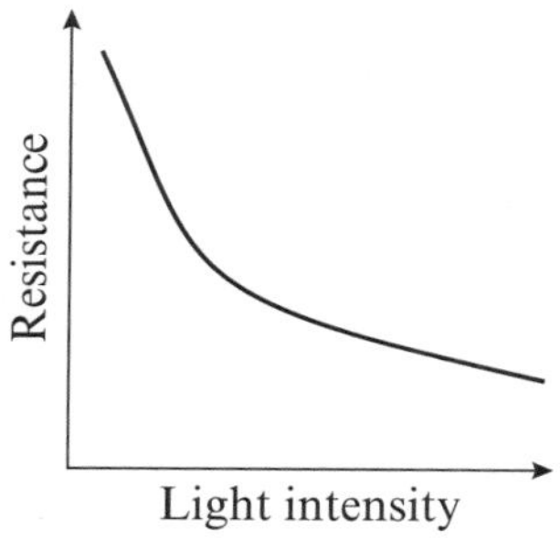

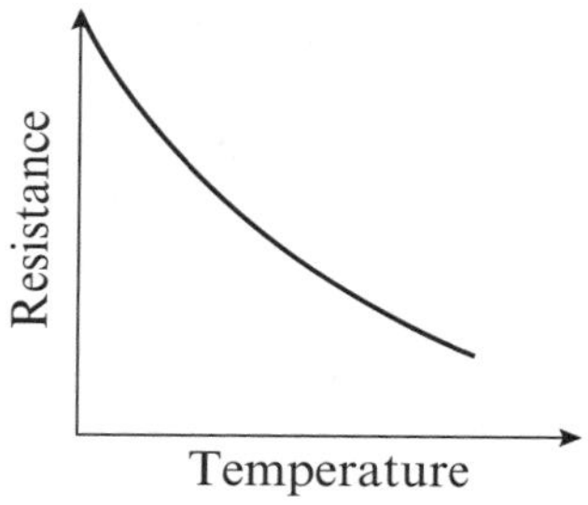

Resistance vs Light intensity (decreasing curve); Resistance vs Temperature (decreasing curve)

.... **(1 mark)**

     (b)   Describe how resistance changes with temperature.

.... **(1 mark)**

4   An electric circuit in a car has a lamp connected in series with the battery and a thermistor. The lamp will only light up when the current is above a certain value. Explain the condition necessary for the lamp to light up.

The lamp lights up when the temperature is ....because the current

through the lamp and the thermistor ....

.... **(2 marks)**

5   The diagram shows a circuit with a light-dependent resistor. State what happens in terms of resistance and current when the level of light increases.

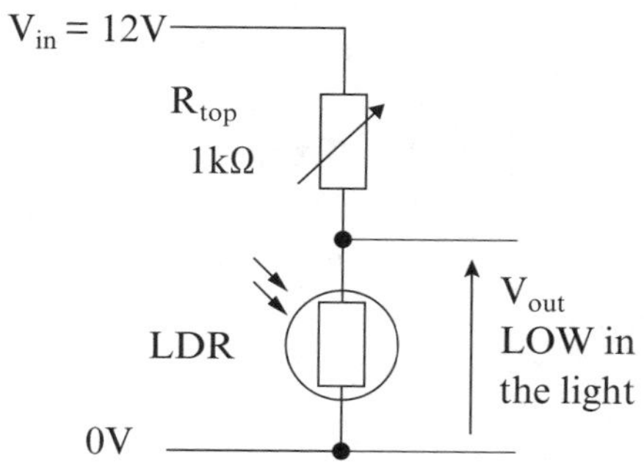

$V_{in} = 12V$
R_{top} $1k\Omega$
LDR
V_{out} LOW in the light
0V

....

....

....

.... **(2 marks)**

Current heating effect

1 Which of the following appliances wastes energy due to the heating effect of a current?

☐ **A** lamp   ☐ **C** electric fire

☐ **B** kettle   ☐ **D** toaster   **(1 mark)**

> **Guided**

2 Explain how the heating effect occurs in a conductor when a potential difference is applied. Include the words in the boxes.

lattice	ions	electrons	collisions

When a conductor is connected to a potential difference ....

....

....

....

....   **(4 marks)**

3 Give three examples of how the heating effect of a current can be used in the home.

....

....

....   **(3 marks)**

4 A student claims that it is safe to use his computer, electric fan heater, desk lamp, coffee maker and toaster from the multi-socket extension lead plugged into a single socket, as all the plugs are earthed. Explain what danger the student is risking.

....

....

....

....

....   **(4 marks)**

5 Traditional filament lamps typically only used 5% of their input energy to produce light. Domestic filament lamps were banned by the European Commission, as part of addressing global energy issues, and replaced with more efficient types of lamps such as LEDs. Explain why traditional filament lamps were considered unsuitable for widespread domestic use and the consequence of using lamps such as LEDs.

> You should refer to the form and approximate amount of wasted energy that was transferred by the filament lamps and how low-energy lamps transfer more useful energy.

....

....

....   **(3 marks)**

Energy and power

1 (a) A hotplate is used to heat a saucepan of water. The hotplate uses mains voltage of 230 V. The electric current through the hotplate is 5 A. Calculate the power of the hotplate in watts.

Using the equation for power P =

power = ... W   (**2 marks**)

(b) A mobile phone has a battery that produces a potential difference of 4 V. When making a call it uses a current of 0.2 A. A student makes a call lasting 30 seconds. Calculate the energy transferred by the mobile phone while the call is made. State the unit.

> You may find the equation
> $E = I \times V \times t$ useful.

energy transferred = ... unit ....   (**3 marks**)

2 The potential difference across a cell is 6 V. The cell delivers 3 W of power to a filament lamp.

(a) Calculate the current flowing through the lamp.

current = ... A   (**3 marks**)

(b) Calculate how much electrical energy is transferred to heat and light in the filament of the lamp when it is switched on for 5 minutes. State the unit.

energy transferred = ... unit ....   (**3 marks**)

(c) The lamp in (b) is replaced by a second lamp with a resistance of 240 Ω which draws a current of 0.5 A. Calculate the power rating of the second lamp.

Using the equation $P = I^2 \times R$

power = ... W   (**3 marks**)

A.c. and d.c. circuits

1 Circuits can operate using either an alternating current or a direct current.

(a) Explain what is meant by an **alternating** current.

an electric current that change direction regularly
And its potential difference ischange   (2 marks)

(b) Explain what is meant by a **direct** current.

Direct current In a wire is the flow of electrons all flow in
the Same direction   (2 marks)

2 Calculate the energy transferred for each of the following appliances:

> **Maths skills** Remember to convert units where appropriate.

(a) a fan heater (2000 W) running for 15 minutes

$$\frac{2000}{900} = 2.2$$

$$E = P \times t_{(s)}$$

energy transferred = $\frac{1800000}{2.2}$ .... J   (2 marks)

(b) a coffee maker (1.5 kW) running for 25 seconds

energy transferred = 5037500 J   (2 marks)

(c) a tablet charger (10 W) running for 6 hours.

$$10 \times 6 \times 60 \times 60$$

energy transferred = 4b216000 J   (2 marks)

3 A cell in an electric circuit causes charged particles to move along the wires as shown in the diagram.

(a) Describe the current supplied by the cell.

The electrons flow round to the
other end of the cell. There must be a complete
circuit for the electrons to flow   (2 marks)

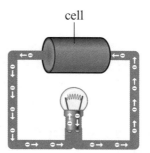

cell

(b) Complete the graph to show what the trace of the current in (a) might look like on an oscilloscope.

Volts

Time

(1 mark)

Mains electricity and the plug

1 (a) Add labels to complete the diagram of a household plug.

4
(4 marks)

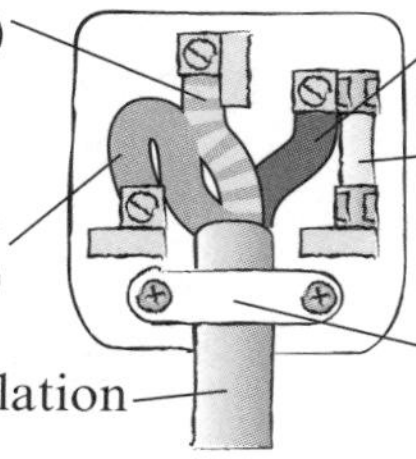

...Earth......
(yellow and green)

...Live......
(brown)

...Fuse...

...Nuehaal...
(blue)

outer insulation

cable grip

(b) Explain which wire the fuse is connected to.

...The fuse is connected to the brown wire....

...

② **(2 marks)**

2 Draw a line from each wire to its correct function.

Wire	Function
brown	Electrical current leaves the appliance at close to 0 V through this wire.
blue	Electrical current enters the appliance at 230 V.
green/yellow	This is a safety feature connected to the metal casing of the appliance.

③ **(3 marks)**

3 Explain how a fuse in a plug works.

> **Guided**

When a ...Current... enters the ...live wire... this produces

...Thermal energy... which ...melts the wire in the fu...

...

The circuit is then ...braku... **(4 marks)**

4 (a) Mains electricity supplies are fitted with a circuit breaker. Explain how a magnetic circuit breaker works.

> Magnetic circuit breakers rely on a high current producing a strong magnetic field.

...If the live wire comes loose The large...
...current heats and melts the wire n the...
...fuse Making an break in the current .② ...

② **(3 marks)**

(b) Describe how the earth wire in a plug protects the user if the live wire becomes loose.

...The earth wire is connected to the Metal casing...
...and a large current flows in though the live...
...wire and out though the earth wire...

③ **(3 marks)**

Extended response – Electricity and circuits

Explain how a circuit can be used to investigate the change in resistance for a thermistor and a light-dependent resistor. Your answer should include a use for each component.

> You will be more successful in extended response questions if you plan your answer before you start writing.
>
> The question asks you to give a detailed explanation of how resistance changes in two types of variable resistor. Think about:
>
> - how resistance in a resistor can be measured and calculated
> - the variable that causes a change in resistance in a thermistor
> - the variable that causes a change in resistance in a light-dependent resistor
> - the consequence to the circuit of a change in resistance in a component
> - uses for thermistors and light-dependent resistors.
>
> You should try to use the information given in the question.

....

....

....

....

....

....

....

....

....

....

....

....

.... **(6 marks)**

Magnets and magnetic fields

1 Complete the diagrams below to show magnetic field lines for:

(a) a bar magnet

(b) a uniform field.

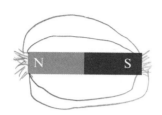

(4 marks)

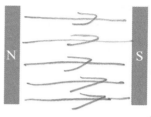

(2 marks)

2 Give two similarities between a bar magnet and the magnetism of the Earth.

Both a bar magnet and the Earth have ..... Magnetic ..fields....................

They also both have similar ....poles ....with ..N.N.and.S............... **(2 marks)**

3 An electric doorbell uses a temporary magnet to move the hammer, and ring the bell, when the button (switch) is pressed.

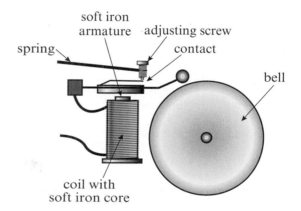

soft iron armature　adjusting screw

spring　　contact

bell

coil with soft iron core

Explain why a temporary magnet rather than a permanent magnet is used for this application.

.....If It was a permanent Magnet then the bell wouldnt work....................................

....

.... **(3 marks)**

4 Rajesh carries out an experiment to test for magnetic or non-magnetic materials by moving a permanent magnet near a range of small objects. Some of the objects are attracted to the magnet. Devise a second test that Rajesh could do to further sort the magnetic objects into temporary and permanent magnets using the same permanent magnet.

> Think of the difference between a temporary and permanent magnet.

.....Put he Materials close to the poll to See If it attracts or repos the object while temporary Magnets Can be Magnetised by permanat So you can See if the objects are Still magnetic When the permanant magnets.............. **(3 marks)**

Had a go ☐   Nearly there ☐   Nailed it! ☐

Current and magnetism

1   The diagram shows a wire passing through a circular card. The **cross** represents the conventional current moving into the page and the **dot** represents the conventional current moving out of the page.

> You can use the 'right-hand grip' rule to help you answer this question.

  (a)   Draw lines of flux on each diagram to show the **pattern** of the magnetic field.   **(2 marks)**

  (b)   Draw arrows on each diagram to show the **direction** of the magnetic field.   **(2 marks)**

2   The magnetic field around a solenoid is similar in shape to the magnetic field of which of these magnets?

     ☐   **A**   ball magnet   ☐   **C**   circular magnet

     ☐   **B**   bar magnet   ☐   **D**   horseshoe magnet   **(1 mark)**

3   (a)   Give two factors that affect the strength of the magnetic field around a current-carrying wire.

> **Guided**

     The strength of the magnetic field depends on ....

     and the ....   **(2 marks)**

  (b)   The graphs below show how the strength of a magnetic field varies with two variables. Label the *x*-axes on the diagrams below with either 'Current' or 'Distance from the wire' to show how they vary with the strength of a magnetic field.   **(2 marks)**

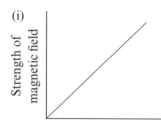

4   In an experiment a student measures the magnetic field strength *B* at a distance of 15 cm from a wire carrying a current of 1.2 A. The experiment is repeated at distances of 7.5 cm and 30 cm.

  (a)   Give the new value of the magnetic field strength in terms of *B* at:

     (i)   7.5 cm from the wire ....   **(1 mark)**

     (ii)   30 cm from the wire. ....   **(1 mark)**

  (b)   The student then changes the distance from the wire to 15 cm to take further measurements of the magnetic field strength but changes the current for the first reading to 0.6 A and for the second reading to 2.4 A. Give the new value of the magnetic field strength for:

     (i)   0.6 A...   **(1 mark)**

     (ii)   2.4 A. ....   **(1 mark)**

Current, magnetism and force

1 A current-carrying wire with a magnetic field is placed in another magnetic field. What is the name of the effect causing the force experienced by the wire?

☐ **A** alternating effect   ☐ **C** induction effect

☐ **B** generator effect   ☐ **D** motor effect   **(1 mark)**

2 (a) Explain the cause of the force experienced by a current-carrying wire near the magnetic field between two magnets.

....

....

.... **(3 marks)**

(b) State when this force is at its maximum.

.... **(1 mark)**

3 Label the diagram showing how Fleming's left hand rule may be used to determine the direction of the force experienced by a current-carrying wire in a magnetic field.

....

....

....

(3 marks)

4 Explain how the size of the force in question **3** may be increased.

The size of the force can be increased by...

....

or by .... **(2 marks)**

5 Calculate the force acting on a wire of length 30 cm carrying a current of 1.4 A when it is placed in a magnetic field of flux density 0.0005 T. Give the unit.

You may find the equation $F = B \times I \times \ell$ useful.

force = ... unit .... **(3 marks)**

Extended response – Magnetism and the motor effect

Devise an experiment to determine the shape and strength of a magnetic field around a long straight conductor.

You will be more successful in extended response questions if you plan your answer before you start writing.

The question asks you to give a detailed explanation of the magnetic field generated by a current-carrying conductor. Think about:

- how you would safely connect the conductor to enable circuit measurements to be taken
- the method you could use to determine the direction of a magnetic field
- the shape of the magnetic field that you would expect to find
- how you would interpret the field patterns of a conductor carrying a current
- how you could change the strength of the magnetic field.

You should use information given in the question.

....

....

....

....

....

....

....

....

....

....

....

....

.... **(6 marks)**

Electromagnetic induction and transformers

1 The diagram shows how a current may be induced in a wire.

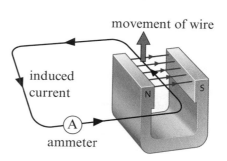

movement of wire

induced current

ammeter

(a) State what must be done to induce a current in the wire that is connected to the ammeter.

.... **(1 mark)**

(b) State how a current in the opposite direction can be induced.

.... **(1 mark)**

(c) State three things that can be done to increase the size of the current.

....

....

.... **(3 marks)**

2 A transformer produces a secondary voltage of 12 V with a current of 1.5 A. The primary voltage is 110 V.

Guided

Calculate the current supplied in the primary coil.

$V_p \times I_p = V_s \times I_s$, $V_p =$.........................., $V_s =$...................., $I_s =$........................

$I_p = V_s \times I_s/V_p =$

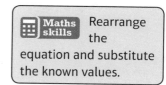
Maths skills Rearrange the equation and substitute the known values.

current in primary coil = ... A   **(3 marks)**

3 Transformers use an effect called electromagnetic induction.

(a) Explain the term electromagnetic induction, as it occurs in a transformer.

....

.... **(2 marks)**

(b) Describe the basic structure of a transformer, including a reason why soft iron is used.

....

....

.... **(3 marks)**

(c) The input voltage to a transformer is 2500 V with an input current of 20 A. The output voltage is 200 000 V. Calculate the output current transferred by the secondary coil.

output current = ... A   **(2 marks)**

Transmitting electricity

1 The National Grid transmits electricity from power stations at 400 000 volts (400 kV). Explain why this voltage is used to transmit electricity long distances.

> Remember that increasing the voltage decreases the current.

....

....

.... **(3 marks)**

2 Identify the correct link between a part of the National Grid and its function.

Part of National Grid	Function
step-down transformer	transmits electrical energy
National Grid system	decreases voltage
power station	increases voltage
step-up transformer	generates electrical energy

(1 mark)

3 A power station generates an electric current of 20 000 A at a voltage of 25 kV. Calculate the power generated in kilowatts.

power generated = ... kW   **(2 marks)**

4 Transformers are used at various places in the National Grid. Describe the role of transformers.

> **Guided**

Step-up transformers ....

as it ....

Near homes, step-down ....

to reduce ....

.... **(4 marks)**

Extended response – Electromagnetic induction

Transformers are used in the transmission of electricity in the National Grid.

Explain how a transformer works, and how and why transformers are used in the National Grid.

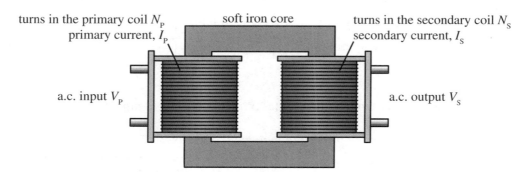

turns in the primary coil N_P
primary current, I_P

soft iron core

turns in the secondary coil N_S
secondary current, I_S

a.c. input V_P

a.c. output V_S

You will be more successful in extended response questions if you plan your answer before you start writing.

The question asks you to give an explanation of how transformers contribute to the National Grid, the supply of electricity, and the role of alternating current and voltage in their use. Think about:

- the meaning of electromagnetic induction
- how a transformer works
- how a transformer can change the size of an alternating voltage
- what step-up and step-down transformers do, and why they are used
- the effect of changing the voltage on the energy lost by heating during transmission.

You should use information given in the question and diagram.

....

....

....

....

....

....

....

....

....

....

....

....

....

.... **(6 marks)**

Changes of state

1 Draw a line from each property to the correct state of matter, and from the state of matter to the correct intermolecular forces. Two lines have been drawn for you.

Property	State	Intermolecular forces
particles move around each other	solid	some intermolecular forces
particles cannot move freely	liquid	almost no intermolecular forces
particles move randomly	gas	strong intermolecular forces

(2 marks)

2 (a) Describe a feature that the three states of matter have in common.

.... **(1 mark)**

(b) Describe two significant differences between the states of matter.

> Different states of matter exist due to differences in energy stores which affect the way the particles behave.

....

.... **(2 marks)**

3 Which statement describes the energy change that takes place when ice melts and then refreezes?

☐ **A** Energy is transferred to surroundings ➔ further energy is transferred to surroundings.

☐ **B** Energy is transferred to the ice ➔ energy is transferred to surroundings.

☐ **C** Energy is transferred to surroundings ➔ energy is transferred to the ice.

☐ **D** Energy is transferred to the ice ➔ energy remains in the system. **(1 mark)**

4 Explain why the temperature stops rising when a liquid is heated to its boiling point and heating continues.

> Guided

At boiling point the ....

....

....

.... **(3 marks)**

5 When water is put into the freezer and turns to ice at 0 °C, explain what happens, in terms of the energy stored, as the temperature continues to fall to −18 °C.

> Remember that the energy stores change as temperature falls.

....

....

....

.... **(3 marks)**

Density

1   A block of wood has a mass of 4000 g and has a volume 5000 cm³. Calculate its density.

density of wood block = ... g/cm³   **(3 marks)**

2   Select the **two** correct statements for density.

☐ **A**   Density is constant for a material at constant temperature.

☐ **B**   Density is related to the atomic packing of a material.

☐ **C**   Density changes with increased mass of a material.

☐ **D**   Density is calculated using force and volume.   **(2 marks)**

3   A metal block measuring 10 cm × 25 cm × 15 cm has a density of 3 g/cm³. Calculate the mass of the block. Give your answer in kilograms.

> **Maths skills**   Work through the calculation first, then convert mass to kilograms at the end.

mass of metal block = ... kg   **(4 marks)**

4   Marco says that all liquids must have lower densities than solids because liquid particles have more kinetic energy and so liquids take up more volume per unit mass than solids. Ella disagrees and says that because solid icebergs float on liquid water, she thinks that Marco must be wrong. Discuss the scientific approaches of these students.

Guided

Marco has approached this problem by ....

....

....

Ella has approached this problem by ....

....

Both students should ....

....   **(5 marks)**

 **Practical skills** **Investigating density**

1 When determining the density of a substance you need to measure the volume of the sample.

(a) State which other quantity you need to measure.

.... **(1 mark)**

(b) Give an example of how you could measure this quantity.

.... **(1 mark)**

2 The volume of a solid object may be determined by two methods.

(a) Describe both methods.

....

.... **(2 marks)**

(b) Explain why one method may be preferable over the other.

....

.... **(2 marks)**

3 (a) Describe the method that can be used to find the density of a liquid.

> **Guided**

Place a measuring cylinder on a ....

....

Add the ....

Record the mass of the... **(3 marks)**

(b) Describe the technique to read the volume of the liquid accurately.

....

.... **(2 marks)**

(c) Calculate the density of a liquid with a mass of 121 g and a volume of 205 cm³. Give the unit.

> You may find this equation useful:
> density = mass ÷ volume   ($\rho = m \div V$)

density = ... unit .... **(2 marks)**

(d) Identify one safety precaution that should be taken when measuring the density of liquids and suggest a method for reducing the risk of harm.

....

....

.... **(2 marks)**

Energy and changes of state

1 State what is meant by the term specific heat capacity.

.... the thermal energy transfered to change the temperature of 1 kg ...... **(1 mark)**

2 Calculate how much energy is required to heat 800 g of water from 30 °C to 80 °C. Take the specific heat capacity of water to be 4200 J/kg °C.

$$\Delta Q = m \times c \times \Delta \theta$$
$$800 \times 4200 \times 50°C = 168\,000\,000 \div 1000 = 168000$$

energy required = ... J   **(3 marks)**

3 Calculate the amount of energy needed to melt 25 kg ice. Take the specific latent heat of fusion of water to be 336 000 J/kg.

$$25 \times 336\,000 = 8\,400\,000$$

energy required = ... J   **(2 marks)**

4 (a) Add the following labels to the graph: melting, boiling.   **(1 mark)**

(b) Explain what is happening at these stages to result in no rise in temperature.

> Consider the bonds between particles.

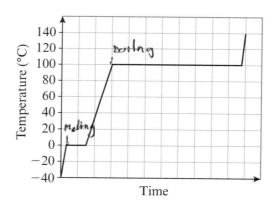

The bonds are getting heated up and are increasing in energy as they change states until all particles have changed to the new state   **(2 marks)**

5 Calculate the specific heat capacity of an 800 g block of metal which is heated for 9 minutes with a current of 2.4 A at 12 V. The temperature increase is 25 °C.

> **Maths skills** Check the units of the quantities before you substitute into the equation.

800 g

$0.8 \times 15552 \times 25 = 311040$

$12 \cup \times 2.4 \times 590 = 15552$

$15552 \div 0.8 \times 25 = 777.6$

specific heat capacity = ... J/kg °C   **(4 marks)**

Thermal properties of water

1   Water is widely used in cooling systems because of its relatively high specific heat capacity compared to some other liquids.

(a) State the definition of the term specific heat capacity.

.... **(1 mark)**

(b) Give the equation for specific heat capacity.

.... **(1 mark)**

2   (a) Describe an experiment that could be set up to measure the specific heat capacity of water using an electric water heater, a beaker and a thermometer.

> **Practical skills** Remember 'pre-experiment' steps e.g. zero the balance to eliminate the mass of apparatus before measuring substances, take a starting temperature reading before heating and decide on the range or type of measurements to be taken.

....

....

....

.... **(5 marks)**

(b) Suggest how you can determine the amount of thermal energy supplied to the heater by the electric current.

....

....

.... **(2 marks)**

(c) Explain how this experiment could be improved to give more accurate results.

....

....

.... **(2 marks)**

3   A known mass of ice is heated until it becomes steam. The temperature is recorded every minute. Describe how to use the data to identify when there are changes of state.

....

....

.... **(2 marks)**

4   Identify two hazards and subsequent safety measures that are common to both experiments to determine specific heat capacity and specific latent heat.

....

....

.... **(2 marks)**

Pressure and temperature

1   Explain what is meant by temperature.

> Consider the movement of particles.

.... **(1 mark)**

2   (a)   Complete the table below showing some equivalent values in kelvin and degrees Celsius.

Kelvin (K)	degrees Celsius (°C)
	0
	−18
373	

(3 marks)

(b)   (i)   Describe what happens to a substance at the temperature absolute zero in terms of pressure and temperature.

....

....

....

....

.... **(3 marks)**

(ii)   Give the value of the Celsius scale at which absolute zero occurs.

absolute zero = ... °C   **(1 mark)**

3   In an experiment a fixed-volume container of 100 g of helium gas is warmed from −10 °C to 30 °C.

(a)   Describe what happens to the velocity of the helium particles as a result of increasing temperature.

As the temperature increases ....

because .... **(2 marks)**

(b)   Explain how this affects the pressure on the container walls.

....

....

.... **(2 marks)**

(c)   State what happens to the average kinetic energy of the particles as the temperature increases.

.... **(1 mark)**

4   In a fixed volume of air the temperature in kelvin is increased by a factor of four. Explain how this affects the average kinetic energy of the air particles.

....

....

.... **(2 marks)**

Extended response – Particle model

Substances can undergo a change of state. Explain, using the kinetic particle theory, the changes of states in water. In your answer, include reasons that explain why thermal energy input and output are not always linked to changes in temperature.

You will be more successful in extended response questions if you plan your answer before you start writing.

The question asks you to give a detailed explanation of how water changes state in terms of particles. Think about:

- the relative kinetic energy of particles in solids, liquids and gases
- why a change of state is described as a reversible change
- how heating the system can result in a change in temperature
- why heating the system does not always result in a change in temperature
- how latent heat is involved in the process.

Include equations to help explain your answer. You should try to use the information given in the question.

....

....

....

....

....

....

....

....

....

....

....

....

.... **(6 marks)**

Elastic and inelastic distortion

1 Draw a line from each force pair to the correct distortion it produces.

Force pair
push forces (towards each other)
pull forces (away from each other)
clockwise and anticlockwise

Distortion
stretching
bending
compression

(3 marks)

2 Give an example where each of the following may occur:

(a) tension

.... **(1 mark)**

(b) compression

.... **(1 mark)**

(c) elastic distortion

.... **(1 mark)**

(d) inelastic distortion.

.... **(1 mark)**

Guided

3 A student investigates loading two aluminium beams each with an elastic limit at 50 N. Beam 1 is tested to 45 N. Beam 2 is tested to 60 N. Predict what you would expect the beams to look like after the experiment. Explain your answer.

mass   aluminium beam

After testing, beam 1 would ....

....

Beam 2 would ....

.... **(4 marks)**

4 Car manufacturers use inelastic distortion to make cars safer. Discuss what they use and what the function of these items is.

> Think about how energy is absorbed to protect the passengers in the event of a crash.

....

....

....

....

.... **(3 marks)**

Springs

1 State what is meant by the term elastic when describing an object experiencing a force.

....

....

.... **(2 marks)**

Guided

2 A spring is stretched from 0.03 m to 0.07 m, within its elastic limit. Calculate the force needed to stretch the spring. Give the unit. Take the spring constant to be 80 N/m.

extension = 0.07 m − ....

force = ... × *extension* ....

....

force = ... unit .... **(3 marks)**

3 Deduce the spring constant that produces an extension of 0.04 m when a mass of 2 kg is suspended from a spring. Take *g* to be 10 N/kg.

> Be careful not to confuse mass with force.

☐ **A** 0.02 N/m

☐ **B** 0.08 N/m

☐ **C** 50 N/m

☐ **D** 500 N/m **(1 mark)**

4 (a) Calculate the spring constant of a spring that is stretched 15 cm when a force of 30 N is applied.

k = ... N/m **(3 marks)**

(b) Calculate the energy transferred to the spring in (a).

> You may find this equation useful:
> *energy* = ½ × *spring constant* × *extension*²

energy transferred = ... J **(2 marks)**

Forces and springs

1 (a) Describe how to set up an experiment to investigate the elastic potential energy stored in a spring using a spring, a ruler, masses or weights, a clamp and a stand.

> Include a step to make sure the spring is not damaged during the experiment.

....

....

....

....

....

.... **(4 marks)**

(b) Explain why it is important to check that the spring is not damaged during the experiment.

....

....

.... **(2 marks)**

(c) Explain how the data collected must be processed before a graph can be plotted. Assume masses are used and measurements made in mm.

Masses must be converted to ....

The extension of the spring must be ....

....

Extension measurements should be .... **(3 marks)**

(d) Describe how a graph plotted from this experiment can be used to calculate:

   (i) the elastic potential energy stored in the spring

     .... **(1 mark)**

   (ii) the spring constant k.

     .... **(1 mark)**

(e) State the name of the law connecting the force, extension and spring constant.

.... **(1 mark)**

(f) Write the equation to calculate the energy stored by the spring.

.... **(1 mark)**

2 Explain the difference between the length of a spring and the extension of a spring.

....

.... **(1 mark)**

Extended response – Forces and matter

Forces can change the shape of an object.

Explain how forces can result in the two types of distortion and give examples of where distortion may be useful.

> You will be more successful in extended response questions if you plan your answer before you start writing.
>
> The question asks you to give a detailed explanation of the how forces acting on an object can cause it to change shape. Think about:
>
> - how forces cause an object to change shape
> - elastic and inelastic distortion
> - how energy may be stored and recovered through change of shape
> - the result of energy not being recovered through change of shape
> - the condition under which metal springs exhibit elastic distortion
> - the relationship between force and extension in the elastic behaviour of springs.
>
> You should use the information given in the question.

....

....

....

....

....

....

....

....

....

....

....

....

....

.... **(6 marks)**

Answers

Biology

1. Plant and animal cells

1 (a) B (**1**)

  (b) C (**1**)

2 (a) Muscles need large amounts of energy for contraction (**1**). This energy is supplied from respiration in mitochondria. (**1**)

  (b) Mitochondria release energy and all cells need energy (**1**), but only leaf (and stem) cells are exposed to light and so have chloroplasts for photosynthesis. (**1**)

3 Cell membrane controls what enters and leaves the cell (**1**); cell wall helps to support the cell / helps it keep its shape. (**1**)

4 Enzymes are proteins and proteins are made on ribosomes (**1**); fat cells don't produce as many proteins as pancreatic exocrine cells. (**1**)

2. Different kinds of cell

1 C (**1**)

2 (a) A = acrosome (**1**); B = flagellum (**1**)

  (b) A contains enzymes to digest a way through the egg cell membrane (**1**); B is used to move the bacterium towards a food source. (**1**)

3 Epithelial cells line tubes (such as trachea) (**1**). Mucus traps dirt / dust / bacteria (**1**) and cilia move mucus along the tubes away from the lungs. (**1**)

4 The egg cell contains nutrients in the cytoplasm (**1**) to supply the growing embryo. (**1**)

3. Microscopes and magnification

1 Light microscopes magnify less than electron microscopes (**1**). The level of cell detail seen with an electron microscope is greater (**1**) because it has a higher resolution. (**1**)

2 (a) because it has a nucleus (**1**) and eukaryotic cells have nuclei (**1**)

  (b) (i) $(23 / 5) \times 2$ (**1**) $= 9.2\,\mu m$ (**1**)

    (ii) $(4 / 5) \times 2$ (**1**) $= 1.6\,\mu m$ (**1**)

  (c) Nuclei are large enough to be seen with a light microscope (**1**) but mitochondria are too small and can only be seen with an electron microscope (**1**) because it has a higher resolution / greater magnification. (**1**)

3 (a) light microscope: $2.5\,\mu m \times 1000$ (**1**) $= 2500\,\mu m$ (or 2.5 mm) (**1**); electron microscope: $2.5\,\mu m \times 100\,000 = 250\,000\,\mu m$ (or 250 mm or 25 cm or 0.25 m) (**1**) (Note that you get the mark for correct use of the formula just once even though you use it twice.)

  (b) The electron microscope (**1**) because it would show more detail / has the correct resolution. (**1**)

4. Dealing with numbers

1 picometre, nanometre, micrometre, millimetre, metre (**1**)

2 5 picometres (**1**), 0.25 grams (**1**), 0.00025 kilograms (**1**), 2500 millimetres (**1**)

3 true (**1**), false (**1**), false (**1**), true (**1**)

4 (a) $0.0309 / 1\,000\,000$ (**1**) $= 3.1 \times 10^{-8}$ m (**1**)

  (b) $0.163 / 250\,000$ (**1**) $= 6.5 \times 10^{-7}$ m (**1**)

  (c) $0.0078 / 800$ (**1**) $= 9.8 \times 10^{-6}$ m (**1**)

5. Using a light microscope

1 (a) (i) to reflect light through the slide (**1**)

    (ii) to hold the slide in place (**1**)

    (iii) to move the objective up and down a long way (**1**)

  (b) (i) because it could crash into the slide (**1**)

    (ii) because it could permanently damage eyesight (**1**)

  (c) (i) a desk / bench / built-in lamp (**1**)

    (ii) Two from: always start with the lowest power objective under the eyepiece (**1**); clip the slide securely on the stage (**1**); move the slide so the cell you need is in the middle of the (low power) view (**1**); use only the fine focusing wheel with the high power objective (**1**)

2 Three from: go back to using the low power objective (**1**); find the part you need and bring it back to the centre view (**1**); focus on it with the coarse focus (**1**); return to the high power objective (**1**) and use the fine focus wheel to bring the part into focus. (**1**)

6. Drawing labelled diagrams

1 (a) Three from: the drawing is in pen rather than pencil (**1**); the title is incomplete (**1**); the magnification is not given (**1**); label lines are not drawn with a ruler (**1**) and cross each other (**1**); not enough cells are shown (**1**) and they are not drawn to scale (**1**); shading should not be used (**1**); lines have been crossed out rather than rubbed out (**1**) and are ragged rather than clear (**1**); the cell membrane can't be seen with the light microscope (**1**)

  (b) Clear drawing of all / most of the cells (**1**); cells not of interest drawn just as outlines (**1**); detail of representative sample of cells (**1**); and avoidance of mistakes from 1 (a) (**1**)

2 Width of image = 45 mm (**1**) so magnification = 45 / 0.113 (**1**) = ×398 (or 400) (**1**)

7. Enzymes

1 The shape of the active site of invertase matches the shape of sucrose but not lactose (**1**), so invertase cannot combine with lactose and catalyse its digestion. (**1**)

2 (a) The rate rises gradually at first (**1**) then reaches a peak at about 40 °C (**1**) and then drops rapidly. (**1**)

  (b) (i) At lower temperatures the molecules move slowly (**1**) so substrate molecules take longer to fit into the active site and react. (**1**)

    (ii) At the optimum temperature (**1**) the enzyme is working at its fastest rate. (**1**)

    (iii) Two from: higher temperatures cause the active site to change shape (**1**) so it can't hold the substrate as tightly (**1**) / the active site breaks up (**1**) and the enzyme is denatured. (**1**)

3 Amylase digests starch in the mouth but is denatured in the stomach (**1**); pepsin has a pH optimum ~2 so digests proteins in the stomach (**1**); pancreatic juice neutralises stomach acidity (**1**); so trypsin and amylase will work in the small intestine. (**1**)

8. pH and enzyme activity

1 (a)

pH	2	4	6	8	10
Time (min)	> 10	7.5	3.6	1.2	8.3
Rate (min)	O	0.13	**0.28**	**0.83**	**0.12**

(**2 marks** for all 5 correct, **1 mark** for 3 correct)

(b)

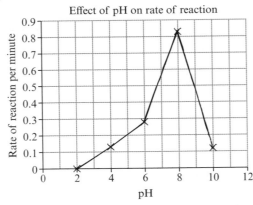

Effect of pH on rate of reaction

correctly labelled axes (**1**), appropriate use of paper (plotted points filling >50% of space) (**1**), correctly plotted points (**1**), line of best fit (can be smooth curve) (**1**)

(c) Two from: use a water bath to control temperature (**1**); repeat several times and take a mean (**1**); use a more accurate method to determine if the film is clear (**1**); use more intermediate pH values (**1**)

9. The importance of enzymes

1

Enzyme	Digests	Product(s)
amylase	starch	**sugars / maltose**
lipase	**lipids**	**fatty acids and glycerol**
protease	**proteins**	amino acids

1 mark for each correct row.

2 (a) Many different enzymes are needed because they are specific for different food molecules (**1**); digestion breaks down the food molecules into molecules small enough to be absorbed. (**1**)

(b) Synthesis reactions occur too slowly (**1**); enzymes are biological catalysts and speed up reactions. (**1**)

3 (a) protease (**1**); needed to break down egg stain, which is made from protein (**1**)

(b) The enzyme is denatured / active site destroyed (**1**) at higher temperatures (**1**), so it would not digest stains as well / would be less active. (**1**)

10. Getting in and out of cells

1 Movement of particles (**1**) from high concentration to low concentration / down a concentration gradient. (**1**)

2 Both occur across a partially permeable membrane / involve movement of molecules. (**1**)

Active transport requires energy / moves molecules against a concentration gradient /
movement from low to high concentration whereas diffusion is passive. (**1**)

3 (a) Osmosis is the net movement of water molecules (**1**) across a partially permeable membrane (**1**) from a low solute concentration (**1**) to a high solute concentration. (**1**)

(b) Gas particles move down the concentration gradient (**1**); there is high concentration of oxygen in the air, low concentration in blood (**1**) and low concentration of carbon dioxide in the air, high concentration in the blood. (**1**)

(c) Glucose must be moved against a concentration gradient (**1**) by active transport that requires energy. (**1**)

11. Osmosis in potatoes

1 Four from: Cut pieces of potato, making sure size / length is the same (**1**); measure mass (**1**); leave in solution for 20 minutes / same time (**1**). Remove from the solution, then measure mass again (**1**). Blot dry before each weighing. (**1**)

2 (a)

Sucrose concentration (mol per dm^{-3})	Initial mass (g)	Final mass (g)	Change in mass (g)	Percentage change (%)
0.0	19.15	21.60	2.45	12.8
0.1	18.30	19.25	0.95	5.2
0.2	15.32	14.85	−0.47	−3.1
0.3	16.30	14.40	−1.90	−11.7
0.5	18.25	16.00	−2.25	−12.3
1.0	19.50	17.20	−2.30	−11.8

Use of correct method (change in mass / initial mass) × 100 (**1**), all six correct. (**1**)

(b) (i)

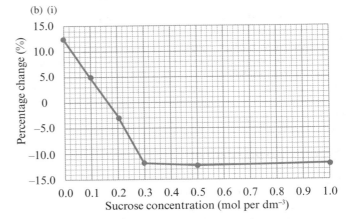

correctly labelled axes (**1**), correctly plotted points (**1**), line of best fit (**1**)

(ii) Solute concentration in range more like 0.16–0.17 mol per dm^{-3} (**1**)

12. Extended response – Key concepts

*Answer could include the following points: (**6**)

- The dye could enter by diffusion or by active transport.
- Active transport acts against a concentration gradient.
- Removal of all the dye from the solutions shows it is working against a concentration gradient, otherwise the solution would still contain some dye.
- Active transport involves enzymes.
- Enzymes work more slowly at lower temperatures.
- Enzymes are destroyed / denatured at high temperature.
- Further experiments: observe cells under light microscope to see if the dye is inside the cells, repeat experiment at different pH values.

13. Mitosis

1 (a) B (**1**)

(b) interphase, prophase, metaphase, anaphase, telophase (**1**)

2 To start with there is 1 cell; after 1 hour this divides into 2 cells. After 2 hours 4 cells. After 3 hours 8 cells. After 4 hours 16 cells. (**1**)

3 (a) A = anaphase (**1**); B = metaphase (**1**)

(b) A = because chromatids are being pulled to each pole (**1**); B = because chromosomes are lined up along the middle of the cell (**1**)

(c) The nuclei divide by mitosis (**1**) but the two new cells do not separate from one another. (**1**)

14. Cell growth and differentiation

1 (a) zygote (**1**)

(b) mitosis (**1**)

2 (a) meristem / root tip / shoot tip (**1**)

(b) Vacuoles take in water by osmosis (**1**) and this causes the cell to elongate. (**1**)

3 (a)

Type of specialised cell	Animal or plant
sperm	animal
xylem	**plant**
ciliated cell	**animal**
root hair cell	**plant**
egg cell	**animal**

3 marks for 4 or 5 correct, **2 marks** for 3 correct, **1 mark** for 2 correct.

(b) Plants: mesophyll cell / guard cell / phloem **(1)**. Animals: small intestine cell / hepatocyte / red blood cell / nerve cell / bone cell / (smooth) muscle cell. **(1)**

4 (a) become specialised **(1)** to perform a particular function **(1)**

(b) because there are many different kinds of specialised cells **(1)** that can carry out different processes more effectively **(1)**

15. Growth and percentile charts

1 (a) C **(1)**

(b) 47.5 − 46.0 **(1)** = 1.5 cm (± 0.2 cm) **(1)**

2 (a) 15.35 − 12.75 = 2.60 g **(1)**;
(2.60 / 12.75) × 100 = 20.4% **(1)**

(b) Any suitable, such as: height **(1)**, measured with a ruler ensuring the stem is vertical **(1)**; shoots / leaves **(1)** by counting number **(1)**

16. Stem cells

1 (a) All the cells in an embryo are stem cells, but in an adult, stem cells are found only in some tissues such as bone marrow **(1)**. Embryonic stem cells can differentiate into many cell types / all diploid cells, but adult stem cells can only differentiate into one type of cell / limited types of cell. **(1)**

(b) (i) meristem **(1)**

(ii) tips of root **(1)** and tips of shoot **(1)**

2 (a) to replace damaged / worn out cells **(1)**

(b) Differentiated cells cannot divide / embryonic stem cells can divide, to produce other kinds of cell. **(1)**

3 (a) Embryonic stem cells could be stimulated to produce nerve cells **(1)** then transplanted into the patient's brain. **(1)**

(b) (i) advantage: easy to extract / can differentiate into nerve cells **(1)**; disadvantage: requires destruction of embryo / may be rejected / may cause cancer **(1)**

(ii) advantage: does not destroy embryos / will not be rejected **(1)**; disadvantage: may cause cancer / may not differentiate into nerve cells **(1)**

17. Neurones

1 (a) Dendrons carry electrical impulses towards the cell body. **(1)**

(b) Axons carry electrical impulses away from the cell body. **(1)**

2 A, axon endings; B, axon; C, cell body; D, dendron; E, myelin sheath; F, receptor cells (in skin) (all correct, **3 marks**; 4 or 5 correct, **2 marks**; 2 or 3 correct, **1 mark**)

3 (a) The cell body of a sensory neurone is in the middle of the axon **(1)**. The cell body of a motor neurone is at the beginning of the neurone. **(1)**

(b) Three from: Dendrites collect impulse from central nervous system **(1)**; axon carries this over long distances through the body **(1)**; nerve ending transmits the impulse to an effector **(1)**; myelin sheath speeds up transmission. **(1)**

4 (a) Myelin sheath speeds up transmission **(1)** because the impulse jumps from one gap to another. **(1)**

(b) Their movement would be impaired / made difficult **(1)** because the nerve impulses to muscles would be slower. **(1)**

18. Responding to stimuli

1 (a) synapse **(1)**

(b) Neurone Y **(1)**; because it is carrying impulses to an effector / muscle **(1)**

(c) When an electrical impulse reaches the end of the neurone X it causes the release of neurotransmitter **(1)** into the gap between the neurones. This substance diffuses **(1)** across the synapse / gap **(1)** and causes neurone Y to generate an electrical impulse. **(1)**

2 (a) stimulus is detected by receptors **(1)**; a nerve impulse travels along a sensory neurone **(1)** then through a relay neurone in the brain / CNS / spinal cord **(1)** and along a motor neurone to an effector. **(1)**

(b) light / movement **(1)** because it causes the eyelid to blink **(1)**

3 Reflex responses are automatic / very fast **(1)** so they protect the body / help avoid danger **(1)** which increases chances of survival. **(1)**

19. Extended response – Cells and control

*Answer could include the following points: **(6)**

- Stages of mitosis described as part of the cell cycle.
- Production of genetically identical daughter cells.
- Diploid number maintained in all cells except gametes.
- Involves replication of DNA.
- Description of cell differentiation.
- Examples of specialised cell types.
- Importance of stem cells: in embryo to produce all different kinds of cell in the body; in adult for growth and repair.

20. Meiosis

1 (a) (i) half the number of chromosomes / one set of chromosomes **(1)**

(ii) sex cells **(1)**

(b) male: sperm **(1)**; female: egg / ovum **(1)**

2 (a) 10 **(1)**

(b) Each daughter cell has only half of chromosomes / genes / DNA from the parent cell **(1)**, so each daughter cell has different combinations of chromosomes / genes / DNA. **(1)**

3 (a) DNA replication **(1)**

(b)

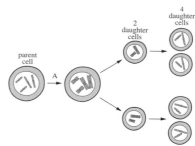

2 daughter cells, then 4 daughter cells **(1)**, one of each pair consisting of duplicated chromosomes in 2 daughter cells **(1)**, 4 daughter cells contain one copy of each pair **(1)**

4 Mitosis maintains the diploid number **(1)** and produces cells that are identical to the parent cell **(1)**. It is used for growth **(1)**. Meiosis creates gametes that have half the number of chromosomes **(1)**. Fertilisation restores the diploid number. **(1)**

21. DNA

1 (a) genome **(1)**

(b) A chromosome consists of a long molecule of DNA packed with proteins **(1)**; a gene is a section of DNA molecule / section of chromosome that codes for a specific protein **(1)**; DNA is the molecule containing genetic information that forms part of chromosomes. **(1)**

2 (a) double helix **(1)**

(b) (i) four **(1)**

(ii) weak hydrogen bonds between complementary bases **(1)**

3 (a) The structure consists of repeated nucleotides / monomers. **(1)**

(b) A base **(1)**, B sugar / ribose **(1)**, C phosphate **(1)**

4 (a) TACCCG **(1)**

(b) because there are complementary base pairs **(1)**; A always pairs with T, C with G **(1)**

Answers

22. Genetic terms

1 (a) (i) different forms of the same gene that produce different variations of the characteristic **(1)**

   (ii) Genotype shows the alleles that are present in the individual, e.g. Bb or BB **(1)**, whereas phenotype means the characteristics that are produced, e.g. brown eyes or blue eyes. **(1)**

   (b) bb **(1)**, BB **(1)**, Bb **(1)**

   (c) bb **(1)** because to have blue eyes she must have two recessive alleles **(1)**

2 There are two copies of each chromosome in body cells **(1)**; each copy has the same genes in the same order **(1)**; a gene is a short piece of DNA at a point on a chromosome **(1)**; genes come in different forms called alleles that produce different variations of the characteristic. **(1)**

23. Monohybrid inheritance

1 (a) correct gametes **(1)**; correct genotypes **(1)**

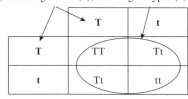

   (b) 25% of the offspring from this cross will be short **(1)**. I know this because tt is short **(1)** and one in four of the possible offspring are tt. **(1)**

   (c) $\frac{3}{4}$ / 75% **(1)**

2 correct parent genotypes Gg **(1)** and gg **(1)**; correct gametes G, g, g, and g **(1)**; correct offspring genotype Gg, gg, Gg, gg **(1)**

24. Family pedigrees

1 C **(1)**

2 (a) two **(1)**

   (b) one **(1)**

   (c) Person 4 does not have cystic fibrosis. This means that they must have one dominant allele from their father **(1)**. But they must have inherited a recessive allele from their mother **(1)**. This means that their genotype is Ff. **(1)**

   (d) Two healthy parents (person 3 & person 4) **(1)** produce a child (person 8) with CF. **(1)**

25. Sex determination

1 (a) X **(1)**; the girl has two X chromosomes, one from each parent **(1)**

   (b) (i) **1 mark** for parental sex chromosomes and **1 mark** for all possible children's chromosomes

	Father	
	X	Y
Mother X	XX	XY
Mother X	XX	XY

   (ii) female **(1)**

2 (a) 50% / ½ / 0.5 **(1)**; depends on which sperm fertilises the egg **(1)** as half the sperm will carry a male sex chromosome / Y chromosome and half the sperm will carry a female sex chromosome / X chromosome. **(1)**

   (b) The statement is not correct **(1)**; the probability of having a child who is a boy is always 50%. **(1)**

26. Variation and mutation

1 (a) Students in a year 7 class will show differences in mass caused by genetic variation **(1)** as well as environmental variation. **(1)**

   (b) Identical twins will only show differences caused by environmental variation. **(1)**

2 (a) Mean height = (181 + 184 + 178 + 190 + 193 + 179) / 6 **(1)** = 184.2 cm **(1)**

   (b) Four from: height is determined partly by genetic factors **(1)** and partly by environmental factors **(1)** such as nutrition **(1)**; different children have inherited different alleles from their parents **(1)**; parents have different heights so will pass on different alleles for height **(1)**; the mean height of the children is greater than that of the parents because of better nutrition / they take after their father more than their mother **(1)**

3 (a) Yellow leaves were caused by environmental factors / are acquired characteristics **(1)** that are not inherited **(1)** but the flower colour was controlled by genes. **(1)**

   (b) A change in the base sequence of DNA **(1)** leads to a large change in the amino acid sequence of the protein **(1)** so that it has a different structure to the normal protein. **(1)**

27. The Human Genome Project

1 (a) the sequence of bases on all human chromosomes **(1)**

   (b) Advantages, two from: a person at risk from a genetic condition will be alerted **(1)**; distinguishing between different forms of disease **(1)**; tailoring treatments for some diseases to the individual **(1)**

   Disadvantages: people at risk of some diseases may have to pay more for life insurance **(1)**; it may not be helpful to tell someone they are at risk from an incurable disease **(1)**

2 Advantages, any two from: she could have earlier / more frequent screening for breast cancer **(1)**; she could consider surgery to remove the breast / mastectomy **(1)**; her doctor might prescribe drugs to reduce the risk of developing cancer **(1)**. Disadvantages, any two from: it might make her more worried / anxious **(1)**; just because she has the mutation doesn't mean she will develop breast cancer **(1)**; she could have unnecessary surgery / medication. **(1)**

28. Extended response – Genetics

*Answer could include the following points: **(6)**

- Mutation is a change in the DNA sequence.
- Most mutations do not affect the phenotype / some mutations have a small effect on the phenotype.
- A single mutation can (rarely) significantly affect the phenotype.
- Human Genome Project maps the DNA / base pairs in the human genome.
- Human Genome Project helps to identify the gene for p53
- This can be linked to the risk of developing cancer.
- Different mutations might increase the risk by different amounts.
- So the risk / whether they carry a harmful allele for a particular individual can be estimated.

29. Evolution

1 (a) (i) Different individuals have different characteristics (because they have different genes / alleles). **(1)**

   (ii) Changes in conditions such as change in availability of food / shelter; change in climate; new predator; new disease. **(1)**

   (b) Because variation in some individuals makes them better at coping with change **(1)** and so more likely to survive **(1)**

2 There is variation within a species **(1)**; members of the species that are most adapted will survive / those that are less well adapted die. **(1)**

3 It will help to classify the new species **(1)** and to find out which other organisms the new species is related to. **(1)**

4 (a) There is variation in the amount of antibiotic resistance in a population of bacteria **(1)**; the most resistant take the longest to die **(1)**, so stopping early means the most resistant will survive and reproduce **(1)** so that all the new population of bacteria will be resistant. **(1)**

   (b) Use of antibiotics is a form of natural selection **(1)**, where only the bacteria with advantageous variation / antibiotic resistance **(1)** will survive and pass these genes on to the next generation. **(1)**

30. Human evolution

1 Three from: toe arrangement **(1)**, length of arms **(1)**, brain size **(1)**, skull shape **(1)**

2 (a) Two from: The older the species, the smaller its brain volume **(1)**; negative correlation / as years before present became less, brain volume increased OR positive correlation – as time 'increases' brain volume increases **(1)**; greatest increase in brain volume between 2.4 and 1.8 million years ago **(1)**; increase in brain volume not linear, increased by $500\,cm^3$ in 2.6 million years **(1)**

(b) an increase in brain volume / size **(1)** to at least $550\,cm^3$ **(1)**

3 (a) The ages of the rock layers where a tool is found can be dated **(1)** by measuring the amount of radiation in the layers. **(1)**

(b) Three from: smooth area in palm of hand **(1)**, will not cut / damage hand **(1)**; chipped section away from hand **(1)** (as it) has sharp edges **(1)** for cutting / unlike smooth area. **(1)**

31. Classification

1 Both have limbs with five fingers **(1)** that have evolved / become adapted to different uses. **(1)**

2 Plants are autotrophic feeders, while animals are heterotrophic feeders **(1)**. Plant cells have cell walls, but animal cells do not have cell walls **(1)**. *You could also say:* Plant cells contain chlorophyll, but animal cells do not. **(1)**

3 Panther / *Panthera pardus* and wolf / *Canis lupus* **(1)**; because they both belong to the same (kingdom, phylum, class and) order **(1)**

4

Domain	Distinguishing characteristic of the domain
Archaea	**cells with no nucleus, genes contain unused sections of DNA**
Eubacteria	**cells with no nucleus, no unused sections in genes**
Eukarya	cells with a nucleus, unused sections in genes

(1 mark for each correct row)

32. Selective breeding

1 (a) Plants or animals with certain desirable characteristics are chosen to breed together **(1)** so that their offspring will inherit these characteristics. **(1)**

(b) pigs with lower body fat are crossed **(1)**; offspring with low body fat are selected and crossed **(1)**; repeated for many generations until a lean breed is produced **(1)**

2 (a) high yield so can feed more people **(1)**; low fertiliser requirement so no need to apply fertiliser / reduce cost **(1)**; pest resistant (or example given) so less pest damage / do not need to apply pesticide **(1)**

(b) drought resistant to cope with times of water shortage without dying **(1)**; tolerant of high temperature **(1)**

(c) less likely to be blown over in the wind / less likely to snap / plant uses less energy in growing the stem, so has more to use for making seeds **(1)**

3 Three from: alleles that might be useful in the future might no longer be available **(1)**; a new disease might affect all organisms and a resistant allele may no longer be available **(1)**; selectively bred organisms might not adapt to changes in climate **(1)**; animal welfare might be harmed **(1)**

33. Genetic engineering

1 Mice do not normally glow, but glow mice have a gene from jellyfish inserted into their DNA / genome **(1)** that produces / codes for a protein that glows in blue light. **(1)**

2 (a) Rice / tomatoes / wheat **(1)**

(b) Two from: increased yields **(1)**; increased nutritional value **(1)**; resistance to attack by insects **(1)**; resistance to herbicides (so that weeds are killed but not the crop) **(1)**; resistance to drought **(1)**

(c) Disadvantage **(1)** with reason **(1)**, e.g. may kill insect species other than pests, so loss of diversity; insects are food for birds, so less food for birds; gene may transfer to another plant (such as a weed) so more of these plants grow amongst the crop plants.

3 Four from: Advantages: can manufacture very pure product **(1)**; can manufacture large quantities of insulin **(1)**; as insulin is human insulin, few problems with rejection **(1)**; overcomes need to harvest insulin from animals **(1)**

Disadvantages: ethical issues over modification of organisms using human genes **(1)**; possibility that GM bacteria may prove unsafe / harmful in the long term **(1)**

(1 mark for each valid point. Full marks are only available if at least one advantage and one disadvantage is given.)

34. Stages in genetic engineering

1 (a) a small circle of DNA from bacteria **(1)**

(b) something that carries a new gene into a cell **(1)**

(c) a few unpaired bases at the ends of double-stranded DNA **(1)**

2 (a) The human gene needed is the one for insulin **(1)**. It is needed because it codes for the protein / hormone. **(1)**

(b) Enzymes have two roles – these are to cut the human gene out of the chromosome **(1)** and insert it into the bacterial DNA. **(1)**

(c) The bacteria provide plasmid **(1)** DNA for the process; the bacteria are useful because they produce large quantities of human insulin. **(1)**

3 (a) The pieces of human DNA have sticky ends **(1)** and the plasmids have matching sets of unpaired bases **(1)** so that the bases in the pieces of human DNA pair up with the bases in the plasmids. **(1)**

(b) DNA ligase links the human DNA and plasmid back into a continuous circle **(1)** so that the plasmids can be inserted back into the bacteria. **(1)**

35. Extended response – Genetic modification

*Answer could include the following points: **(6)***

- both rely on variation caused by mutation and sexual reproduction / meiosis
- evolution involves natural selection
- variation means some individuals are better able to survive in their environment
- they will produce more healthy offspring than others
- so their alleles are more likely to be passed on
- also explains how organisms adapt to changes in their environment over several generations
- selective breeding involves artificial selection
- plants or animals with desired characteristics are bred together
- offspring will inherit these characteristics
- those with the most desirable characteristics are bred further
- repeated many times
- does not involve adaptation to changed environment
- but can be used to breed plants / animals better able to cope with difficult conditions.

36. Health and disease

1 (a) being free from disease and eating and sleeping well **(1)**

(b) how you feel about yourself **(1)**

(c) how well you get on with other people **(1)**

2 (a) Communicable: ✓influenza, ✓tuberculosis, ✓*Chlamydia*; Non-communicable: ✓lung cancer, ✓coronary heart disease

(3 marks for 5 correct, **2 marks** for 3 or 4 correct, **1 mark** for 1 or 2 correct)

(b) Communicable: can be transmitted between people **(1)**; non-communicable: cannot be transmitted between people, instead caused e.g. by lifestyle or environmental factors. **(1)**

3 (HIV) causes damage to the immune system **(1)**; reduced immune response / immunity **(1)**

4 (a) Three from: a virus infects a body cell **(1)** and takes over the body cell's DNA **(1)** causing the cell to make toxins **(1)** or damages the cell when new viruses are released. **(1)**

(b) Bacteria can release toxins **(1)** and can invade and destroy body cells. **(1)**

37. Common infections

1 (a) Zimbabwe **(1)**; 15.1–14.3 = 0.8% decrease **(1)**

(b) All countries show a decrease in the % of 15 to 49 year olds with HIV **(1)**, one example of a trend such as: all % have dropped somewhere between 0.3 and 2.9%. **(1)**

2 (a) D **(1)**

(b) Two from: leaf loss; **(1)**; bark damage **(1)**; dieback of top of tree **(1)**

3

Disease	Type of pathogen	Signs of infection
cholera	**bacterium**	watery faeces
tuberculosis	bacterium	persistent cough – may cough up blood
malaria	**protist**	**fever, weakness, chills and sweating**
HIV	**virus**	mild flu-like symptoms at first

(all correct for **3 marks**, 3 correct for **2 marks**, 2 or 1 correct for **1 mark**)

4 (a) bacterium **(1)**

(b) Two from: inflammation in stomach **(1)**; bleeding in stomach **(1)**; stomach pain **(1)**

38. How pathogens spread

1 C **(1)**

2

Disease	Pathogen	Ways to reduce or prevent its spread
Ebola haemorrhagic fever	**Virus (1)**	keep infected people isolated, wear full protective clothing while working with infected people or dead bodies
Tuberculosis	Bacteria	**ventilate buildings to reduce chance of breathing in bacteria / diagnose promptly and give antibiotics to kill bacteria / isolated infected people (1)**

3 Boil water before drinking / wash hands after using toilet **(1)** because bacteria are spread in water / by touch **(1)**

4 (a) The bacteria are spread in water **(1)**; in developed countries water is treated to kill pathogens / good hygiene prevents their spread **(1)**.

(b) to prevent being infected by the Ebola virus **(1)** because Ebola virus is present in body fluids of infected people even after death **(1)**

39. STIs

1 An infection spread by sexual activity **(1)**

2 B **(1)**

3

Mechanism of transmission	Precautions to reduce or prevent STI
unprotected sex with an infected partner	using condoms during sexual intercourse **(1)**
sharing needles with an infected person (1)	supplying intravenous drug abusers with sterile needles
infection from blood products	**screening blood transfusions (1)**

4 (a) Screening helps identify an infection **(1)** so people can be treated for the disease / take extra precautions to prevent transmission. **(1)**

(b) HIV is a virus **(1)** and viruses cannot be treated with antibiotics. **(1)**.

40. Human defences

1 (a) Skin acts a physical barrier that stops microorganisms getting into the body. **(1)**

(b) Hydrochloric acid in the stomach kills pathogens. **(1)**

(c) (i) lysozyme **(1)**

(ii) kills bacteria **(1)** by digesting their cell walls **(1)**

2 (a) (i) mucus **(1)**

(ii) sticky so traps bacteria / pathogens **(1)**

(b) (i) cilium **(1)**

(ii) The cilia on the surface of these cells move in a wave-like motion **(1)** and this moves mucus and trapped pathogens out of the lungs **(1)** towards the back of the throat where it is swallowed. **(1)**

(c) Mucus travels down to into the lungs carrying pathogens **(1)**, because the cilia cannot move and take the pathogens back up to the throat. **(1)**

41. The immune system

1 lymphocytes **(1)**

2 Pathogens have substances called antigens **(1)** on their surface. White blood cells called lymphocytes **(1)** are activated if they have antibodies **(1)** that fit these substances. These cells then divide many times to produce clones / identical cells **(1)**. They produce large amounts of antibodies that stick to the antigens / destroy the pathogen. **(1)**

3 (a) Lymphocytes producing antibodies against measles virus are activated **(1)**; these lymphocytes divide many times **(1)**; so concentration of antibodies increases **(1)** then decreases when the viruses have all been destroyed **(1)**.

(b) Some of the lymphocytes stay in the blood as memory lymphocytes **(1)**; these respond / divide after infection **(1)**; so the number of lymphocytes producing the antibodies against the measles virus increases rapidly **(1)**.

(c) (i) (The girl had not been exposed to the chicken pox virus before because) line B is similar in size and shape to line A **(1)**, which was for a first infection with measles / the line would be higher if its was a second infection. **(1)**

(ii) The concentration of antibodies increased faster / to a higher concentration **(1)**; so the measles viruses were destroyed before they could cause illness / symptom / disease. **(1)**

42. Immunisation

1 (a) Artificial immunity to a pathogen **(1)** by using a vaccine **(1)**

(b) A vaccine contains antigens from a pathogen **(1)**, often in the form of dead / weakened pathogens. **(1)**

(c) The vaccine produces memory lymphocytes **(1)** so if the person is exposed to the disease the memory lymphocytes produce a very rapid secondary response **(1)** so it is very unlikely they will become ill. **(1)**

2 (a) 2003

(b) The number of cases would increase **(1)** because fewer babies were immunised. **(1)**

43. Treating infections

1 (a) C **(1)**

(b) Antibiotics kill bacteria / inhibit their cell processes **(1)** but do not affect human cells. **(1)**

2 The pharmacist's advice would be not to take the penicillin **(1)**. The man's cold is due to a virus, so the penicillin will not be effective in combating the infection. **(1)**.

3 (a) Sinusitis is (probably) not caused by a bacterial infection. **(1)**

(b) Same number of patients got better without antibiotic **(1)** although the patients taking antibiotic may have got better more quickly. **(1)**

44. New medicines

1 (a) 3, 1, 5, 2, 4 (All 5 correct = **3 marks**, 3 correct = **2 marks**, 1 correct = **1 mark**)

(b) (i) testing in cells or tissues to see if the medicine can enter cells and have the desired effect **(1)**; testing on animals to see how it works in a whole body / has no harmful side effects **(1)**

(ii) by testing in a small number of healthy people **(1)**

(c) Medicine is tested on people with the disease that it will be used to treat **(1)** so that the correct dose can be worked out **(1)** and to check for side effects in different people. **(1)**

2 (a) Large number of subjects make the data valid **(1)** and repeatable **(1)**; OR side effects may not be seen in small numbers **(1)** so it is easier to notice side effects with a large trial group **(1)**; OR there are different stages of the trial **(1)** and each step needs a different group of people. **(1)**

(b) The medicine appears to be effective in nearly 400 people with high blood pressure (**1**); this reduction is much greater than those in the placebo group (**1**). You could also say: the medicine seems to have very little adverse effect on the blood pressure of those in the 'normal' group (so it is effective).

45. Non-communicable diseases

1 An infectious disease is caused by a pathogen (**1**) and is passed from one person to another (**1**). A non-communicable disease is not passed from one person to another. (**1**)

2 Three from: inherited / genetic factors (**1**); age (**1**); sex (**1**); ethnic group (**1**); lifestyle (e.g. diet, exercise, alcohol, smoking) (**1**); environmental factors (**1**)

3 (a) (i) Bangladeshi men (**1**)

    (ii) black women (**1**)

(b) Four from: the prevalence of CHD increases with age (**1**); overall the prevalence is higher in men than in women (**1**), but prevalence is similar in black men and women (**1**); Bangladeshi men have the highest prevalence but Bangladeshi women are in the middle (**1**); ethnic group seems to be a bigger factor in men than in women (**1**); the prevalence in all ethnic groups is very similar in the 40–49 age group (**1**)

46. Alcohol and smoking

1 (a) Ethanol is a drug that is toxic / poisonous to cells (**1**). It is broken down by the liver and harms liver cells (**1**). Too much alcohol over a long period causes cirrhosis / liver disease. (**1**)

(b) because it is caused by how we choose to live (**1**)

2 Two from: because carbon monoxide in cigarette smoke (**1**) reduces how much oxygen the blood can carry to the baby (**1**), leading to low birth weight in babies / other abnormalities. (**1**)

3 (a) Two from: cardiovascular disease (**1**); lung cancer (**1**); respiratory / lung disease (**1**)

(b) Substances in cigarette smoke cause blood vessels to narrow (**1**) which increases the blood pressure (**1**) leading to cardiovascular disease. (**1**)

47. Malnutrition and obesity

1 (a) too little of one or some nutrients in the diet (**1**)

(b) Four from: anaemia increases with increasing age (**1**) in both men and women (**1**), but whereas there is an increase in females from 1–16 and 17–49 (**1**) followed by a decline (**1**), in males the lowest age groups are 17–49 and 50–64. (**1**)

2 (a)

Subject	Weight (kg)	Height (m)	BMI
person A	80	1.80	24.7
person B	90	1.65	33.1
person C	95	2.00	23.8

All 3 correct = **2 marks**, 2 correct = **1 mark**

(b) person B (**1**)

3 Too much fat / obesity increases the risk of cardiovascular disease (**1**) and abdominal fat is most closely linked with cardiovascular disease (**1**); measuring waist : hip ratio is a better measure of abdominal fat. (**1**)

48. Cardiovascular disease

1 (a) Two from: lifestyle changes (**1**); medication (**1**); surgery (**1**)

(b) Two from: give up smoking (**1**); take more exercise (**1**); eat a healthier diet (lower fat, sugar and salt) (**1**); lose weight (**1**)

(c) because cardiovascular disease reduces life expectancy (**1**) and can be fatal before treatment can be given (**1**)

2 Lifestyle changes – Benefits: no side effects / may reduce chances of getting other health conditions / the cheapest option.

Drawbacks: may take time to work / may not work effectively.

Medication – Benefits: easier to do than change lifestyle / starts working immediately / cheaper and less risky than surgery. Drawbacks:

can have side effects / needs to be taken long term / may not work well with other medication the person is taking

Surgery – Benefits: once recovered, there are no side effects / usually a long-term solution. Drawbacks: there is a risk the person will not recover after the operation / risk of infection after surgery / expensive / more difficult to do than giving medication

(**3 marks** for 6 correct, **2 marks** for 4 or 5 correct, **1 mark** for 2 or 3 correct)

3 Surgery can help prevent heart attacks / strokes (**1**) but costs more than inserting a stent (**1**) and surgery has more risk (e.g. risk of infection). (**1**). However, it can be a long-term solution / other suitable conclusion. (**1**)

49. Extended response – Health and disease

* Answer could include the following points: (**6**)

- Communicable diseases caused by infection with a pathogen
- Non-communicable diseases are not caused by infection
- Non-communicable diseases may have genetic causes
- Non-communicable diseases may be caused by environmental / lifestyle factors, e.g. diet / nutrition / alcohol / smoking
- But (poor) diet / nutrition can also increase risk of catching infections
- Treatment of communicable diseases largely through medicines, e.g. antibiotics
- Non-communicable diseases can be treated with medicines but also through lifestyle changes.

50. Photosynthesis

1 Plants or algae are photosynthetic organisms / producers (**1**) so they are the main producers of biomass (**1**) and animals have to eat plants / algae. (**1**)

2 (a) carbon dioxide + water (**1**) → glucose + oxygen (**1**)

(b) The product of photosynthesis / glucose has more energy than the reactants (**1**) because energy is transferred from the surroundings / light. (**1**)

3 (a) Light is required for photosynthesis (**1**) because only the leaf exposed to light produced starch. (**1**)

(b) Chlorophyll / chloroplasts required for photosynthesis (**1**) because only green areas / areas with chlorophyll or chloroplasts produced starch. (**1**)

51. Limiting factors

1 (a) temperature (**1**)

(b) Add algal balls to hydrogen carbonate solution (**1**). Leave for a set amount of time e.g. 2 hours (**1**).
Compare the colour change against standard colours (**1**).

2 (a) Increasing the carbon-dioxide concentration increases the rate of photosynthesis. (**1**)

(b) Adding carbon-dioxide means it will no longer be a limiting factor (**1**) so the plants can make more sugars needed for growth. (**1**)

(c) You could the increase the temperature (**1**) as this would make photosynthesis happen faster / more quickly. (**1**)

3 Increased temperature increases the rate of reaction (**1**) so photosynthesis / growth happens faster (**1**); eventually other factors limit rate / rate reaches maximum (**1**); higher temperatures denature enzymes responsible for photosynthesis (**1**)

52. Light intensity

1 (a) points plotted accurately (**1**) curve of best fit drawn (**1**)

(b) 76 (± 2) (**1**)

(c) the greater the light intensity, the higher the rate (**1**); not a linear relationship (**1**)

(d) (i) Take care not to touch the bulb if it is hot. (**1**)

    (ii) Place a water tank next to the bulb if it is hot (**1**) to help prevent heat from the lamp reaching the test tube (**1**); OR use a ruler to make sure that the lamp is at the measured distance (**1**) because differences in distance will change the light intensity (**1**); you could also suggest to repeat the experiment. (**1**)

(e) You could use the light meter to measure the light intensity **(1)** at each distance **(1)** and then plot a graph of rate of photosynthesis bubbling against light intensity. **(1)**

53. Specialised plant cells

1 (a) phloem **(1)**

(b) A   There are holes **(1)** to let liquids flow from one cell to the next. **(1)**

B   There is small amount of cytoplasm **(1)** so there is more space for the central channel. **(1)**

(c) Mitochondria supply energy **(1)** for active transport (of sucrose). **(1)**

2 (a) xylem **(1)**

(b) Three from: the walls are strengthened with lignin rings to prevent them from collapsing **(1)**; no cytoplasm means there is more space for water **(1)**; pits in the walls allow water and mineral ions to move out **(1)**; no end walls means they form a long tube so water flows easily **(1)**

54. Transpiration

1 (a) Transpiration is the loss of water **(1)** by evaporation from the leaf surface. **(1)**

(b) stomata (in the leaf) **(1)**

(c) (i) moves faster **(1)** because a faster rate of water loss from leaves **(1)**

(ii) moves more slowly **(1)**; stomata covered so a lower rate of water loss **(1)**

2 (a) Guard cells take in water by osmosis **(1)** so they swell causing the stoma to open **(1)**; when the guard cells lose water they become flaccid / lose rigidity and the stoma closes. **(1)**

(b) The stomata are open during the day, so water is lost by transpiration **(1)** faster than it can be absorbed by the roots **(1)**. Water is lost from the vacuoles and the plant wilts. At night, the stomata close so water is replaced. **(1)**

55. Translocation

1 (a) the movement of sucrose around a plant **(1)**

(b) A **(1)**

2 (a) Radioactive carbon dioxide is supplied to the leaf **(1)**; radioactive carbon / sucrose will then be detected in the phloem **(1)**; and eventually incorporated into starch in the potato. **(1)**

(b) Radioactivity would remain in the leaf / not appear in the phloem **(1)** because companion cells actively pump sucrose into / out of the phloem. **(1)**

3

Structure or mechanism	Transport of water	Transport of sucrose
Xylem	X	
Phloem		X
Pulled by evaporation from the leaf	X	
Requires energy		X
Transported up and down the plant		X

(**1 mark** for each correct row.)

56. Water uptake in plants

1 (a) Rate of transpiration increases **(1)** because increasing light intensity causes stomata to open. **(1)**

(b) Rate of transpiration increases **(1)** because higher temperature increases energy of water molecules / water molecules move faster. **(1)**

2 (a) The rate of evaporation was higher when the fan was on **(1)**; because the movement of air removes water more quickly from the leaves **(1)**, increasing the concentration gradient from leaf to air. **(1)**

(b) The rate of evaporation became quicker than the rate at which the plant could take up water **(1)**; the stomata of the plant closed **(1)** to prevent evaporation from occurring / conserve water. **(1)**

(c) Volume of 90 mm length of tube = ($\pi \times 0.25^2 \times 90$) ÷ 5 **(1)** = 3.5 mm^3 / min **(1)**

57. Extended response – Plant structures and functions

*Answer could include the following points: **(6)**

- Guard cells can take in water by osmosis, and swell.
- This happens in the light.
- When the guard cells swell, they become rigid and the stomata / stoma opens.
- Carbon dioxide diffuses into the leaf through the stomata and is used in photosynthesis.
- Water is lost from the leaf through open stomata.
- This helps move water and dissolved mineral ions up the stem by transpiration.
- At night, the guard cells lose water and the stomata / stoma closes.
- Closing stomata when it is dark and no photosynthesis is occurring helps reduce water loss.

58. Hormones

1 (a) Hormones are produced by endocrine glands **(1)** and are released into the blood **(1)**. They travel round the body until they reach their target organ **(1)**, which responds by releasing another chemical substance. **(1)**

(b) Hormones have long-lived effects; nerves have short-term effects **(1)**. Nerve impulses act quickly; hormones take longer. **(1)**

2 A = hypothalamus **(1)**, B = pituitary **(1)**, C = thyroid **(1)**, D = pancreas **(1)**, E = adrenal **(1)**, F = testis **(1)** and G = ovary **(1)**

3

Hormone	Produced in	Site of action
TRH	**hypothalamus**	pituitary gland
TSH	pituitary gland	**thyroid gland**
ADH	pituitary gland	**kidney**
FSH and LH	**pituitary gland**	ovaries
insulin and glucagon	**pancreas**	liver, muscle and adipose tissue
adrenalin	**adrenal gland**	various organs, e.g. heart, liver, skin
progesterone	**ovaries**	uterus
testosterone	**testes**	male reproductive organs

All 8 correct = **4 marks**, 6 or 7 correct = **3 marks**, 4 or 5 correct = **2 marks**, 2 or 3 correct = **1 mark**.

59. Adrenalin and thyroxine

1 (a) Two from: heart beats faster **(1)** so oxygen is carried around the body faster **(1)**; OR some blood vessels constrict **(1)** so blood pressure increases **(1)**; OR some blood vessels dilate **(1)** so blood flow to muscles increases / delivery of oxygen and nutrients to muscles increases **(1)**; OR liver converts glycogen to glucose **(1)** so more glucose available for respiration **(1)**

(b) An increase in thyroxine concentration causes changes **(1)** that bring about a decrease in the amount of thyroxine released **(1)** (or vice versa).

(c) When blood concentration of thyroxine is lower than normal **(1)** hypothalamus produces TRH **(1)** which stimulates pituitary to produce TSH **(1)** which stimulates thyroid gland to produce thyroxine **(1)** (or reverse if thyroxine concentration is higher than normal).

2 Thyroxine controls the resting metabolic rate **(1)** so is produced constantly **(1)**. Adrenalin is produced in response to fright / excitement **(1)** and is only produced when needed. **(1)**

60. The menstrual cycle

1 Two from: oestrogen **(1)**; progesterone **(1)**; FSH **(1)**; LH **(1)**

2 (a) A = menstruation **(1)**; B = ovulation **(1)**

(b) any time between day 14 and about day 17 **(1)**

(c) The lining of the uterus breaks down **(1)** and is lost in a bleed or period. **(1)**

3 (a) Pills, implants or injections release hormones that prevent ovulation **(1)**, and thicken mucus at the cervix **(1)** preventing sperm **(1)** from passing.

(b) (i) These figures are maximum values **(1)** when the methods are used correctly. **(1)**

(ii) hormonal method most effective **(1)**; diaphragm / cap least effective **(1)**; only condom offers protection against STIs **(1)**

61. Control of the menstrual cycle

1 (a) pituitary gland **(1)**, ovary / mature follicle **(1)**

(b) corpus luteum **(1)**, pituitary gland / lining of uterus **(1)**

(c) ovary / maturing follicle **(1)**, the pituitary gland **(1)**

2 (a) (i) Levels of progesterone are low **(1)**, which allows FSH release by pituitary gland. **(1)**

(ii) Level of LH rises rapidly / LH surge **(1)** triggers ovulation. **(1)**

(iii) Progesterone levels fall after day 23 **(1)** so uterus wall thickness is not maintained and therefore pregnancy has not occurred. **(1)**

(b) High level of progesterone **(1)** prevents release of FSH **(1)** so follicles aren't stimulated to grow. **(1)**

(c) Increase in temperature coincides with ovulation **(1)** so she will be more fertile in the days after. **(1)**

62. Assisted Reproductive Therapy

1 (a) Clomifene helps increase the concentration of FSH and LH in the blood **(1)**, so stimulates the maturation of follicles and release of eggs. **(1)**

(b) Clomifene will stimulate the release of eggs **(1)**, but these cannot reach the uterus if the oviducts are blocked. **(1)**

2 (a) FSH stimulates maturation of follicles **(1)**; FSH stimulates release of many eggs. **(1)**

(b) eggs and sperm mixed to allow fertilisation **(1)**; one or two healthy embryos placed in uterus **(1)**

(c) Cell can be removed from embryo before being placed in the uterus **(1)**; this can be tested for genetic disorders **(1)**.

3 Four from the following, but there must be at least one drawback and at least one advantage.

Drawbacks: the success rate for IVF is still quite low **(1)**; at only 12 400 / 45 250 × 100 = 27% **(1)**; cost is quite high (especially if more than one cycle of treatment is needed) **(1)**; total cost to the NHS in 2010 = 45 250 × £2500 = £113 million **(1)**

Benefits: it allows couples to have children if they are not able to conceive naturally **(1)**; even though success rate is low, it is still successful for some couples **(1)**; it can be easier to use this procedure than to adopt **(1)**

63. Blood glucose regulation

1 order is: 3, 5, 1, 4, 2. All 5 correct = **3 marks**, 3 correct = **2 marks**, 1 correct = **1 mark**.

2 (a) A&F pancreas **(1)**, B insulin **(1)**, C&D liver **(1)**, E glucagon **(1)**

(b) (i) Insulin causes the liver to take up excess glucose and convert it to glycogen **(1)**, causing blood glucose concentration to fall. **(1)**

(ii) Glucagon causes the liver to convert glycogen to glucose, which is released into the blood **(1)**, causing the blood glucose concentration to rise. **(1)**

64. Diabetes

1 (a) As the BMI increases the percentage of people with diabetes increases **(1)** so there is a positive correlation. **(1)**

(b) (i) BMI = 88 ÷ 1.8² = 27.2 **(1)**; he (is overweight so) has an increased risk of Type 2 diabetes **(1)**, but not the highest risk. **(1)**

(ii) waist : hip ratio = 104 ÷ 102 = 1.02 which is obese **(1)**, so he has a high risk of developing Type 2 diabetes **(1)**, because there is a correlation between W:H ratio and risk of Type 2 diabetes. **(1)**

2 (a) Controlling diets will help to control the number of people who are obese **(1)**. Fewer obese people means fewer people with diabetes. **(1)**

(b) (i) In Type 1 diabetes, no insulin is produced so it has to be replaced with injections **(1)**, but in Type 2 diabetes, organs don't respond to insulin. **(1)**

(ii) because a large meal means a higher blood glucose concentration **(1)** so more insulin is needed to reduce the glucose concentration **(1)**

65. Extended response – Control and coordination

*Answer could include the following points: **(6)**

- Cause of Type 1 diabetes: immune system has damaged insulin-secreting cells in pancreas, so no insulin produced.
- Cause of Type 2 diabetes: insulin-releasing cells may produce less insulin and target organs are resistant / less sensitive to insulin.
- Link risk of Type 2 diabetes with obesity / BMI / waist : hip ratio.
- Treat Type 1 diabetes by injecting insulin. Amount of insulin injected can be changed according to the blood glucose concentration.
- Treat Type 2 diabetes by diet (eating healthily and reduced sugar) and exercise.
- Treat more severe Type 2 diabetes with medicines to reduce the amount of glucose the liver releases or to make target organs more sensitive to insulin.

66. Exchanging materials

1 (a) kidneys / nephrons **(1)**; to maintain constant water level / osmoregulation **(1)**

(b) kidneys / nephrons **(1)**; urea is a toxic waste product **(1)**

2 small intestine **(1)**; because they are needed for energy / respiration / growth / repair **(1)**

3 lungs **(1)**; oxygen is needed for respiration **(1)**; carbon dioxide is a waste product **(1)**

4 (a) The surface of the small intestine is covered with villi **(1)**. These help by increasing the surface area available for absorption. **(1)**

(b) This makes the absorption of food molecules more efficient / effective **(1)** by reducing the distance that the molecules have to diffuse. **(1)**

5 Four from: The flatworm is very flat and thin **(1)** which means it has a large surface area:volume ratio **(1)**; the earthworm is cylindrical so has smaller surface area:volume ratio **(1)**; every cell in the flatworm is close to the surface **(1)**; in the earthworm diffusion has to happen over too great a distance / through too many layers of cells. **(1)**

67. Alveoli

1 (a) Oxygen diffuses from the air in alveoli into the blood in capillaries **(1)**. Carbon dioxide diffuses from the blood into the air. **(1)**

(b) Millions of alveoli create a large surface area for the diffusion of gases **(1)**. Each alveolus is closely associated with a capillary **(1)**. Their walls are one cell thick **(1)**. This minimises the diffusion distance **(1)**.

2 maintains concentration gradient **(1)** which maximises the rate of diffusion **(1)**

3 Three from: breathlessness / shortness of breath / similar **(1)**; less oxygen in blood than normal **(1)** so less respiration / energy **(1)**; increased carbon dioxide concentration reduces pH **(1)** which affects enzyme-controlled reactions **(1)**

68. Blood

1 (a) Red blood cells carry oxygen around the body **(1)** and haemoglobin binds to oxygen. **(1)**

(b) Biconcave shape increases surface area **(1)** to allow more diffusion of oxygen **(1)**; no nucleus **(1)** allows more space for haemoglobin **(1)**.

2 Dissolved substances such as glucose / oxygen are transported to tissues **(1)** where they are used in respiration **(1)**. *You could also say* Waste such as urea / carbon dioxide **(1)** is transported away from tissues, to kidney / lungs. **(1)**

3 Platelets respond to a wound by triggering the clotting process **(1)**; the clot blocks the wound **(1)** and prevents pathogens from entering. **(1)**

4 (a) Infections are caused by pathogens **(1)**; lymphocytes produce antibodies **(1)** that stick to them and destroy them. **(1)**

(b) Phagocytes surround foreign cells **(1)** and digest them. **(1)**

Answers

69. Blood vessels

1 (a) An artery has thick walls (**1**). These walls are composed of two types of fibres: connective tissue (**1**) and elastic fibres. (**1**)

(b) Wall stretches as blood pressure rises / heart ventricles contract (**1**) and recoils (*not contracts!*) when blood pressure falls / heart ventricles relax. (**1**)

2 (a) Thin walls / only one cell thick (**1**) run close to almost every cell. (**1**)

(b) faster diffusion of substances (**1**); because of short distance / large surface area (**1**)

3 (a) (i) Blood flows at low pressure (**1**) so no need for elastic wall of arteries / need wide tube in veins. (**1**)

(ii) Muscles contract and press on veins (**1**); blood is pushed towards the heart because valves prevent flow the wrong way. (**1**)

(b) Veins have a thinner muscle wall than arteries (**1**) so it is easier to get the needle in (**1**). OR Veins contain blood under lower pressure (**1**) so taking blood is more controlled. (**1**)

70. The heart

1 aorta – carries blood from heart to body (**1**); pulmonary artery – carries blood from heart to lungs (**1**); pulmonary vein – carries blood from lungs to heart (**1**); vena cava – carries blood from body to heart (**1**)

2 (a) because it acts as a pump (**1**) and muscles contract to pump the blood (**1**)

(b) order of parts: (vena cava) right atrium, right ventricle, pulmonary artery (lungs), pulmonary vein, left atrium, left ventricle (aorta) (names all correct for **2 marks**, 4 correct for **1 mark; additional mark** for correct order)

3 (a) right ventricle (**1**); pumps blood to the lungs / pulmonary artery (**1**)

(b) heart valve closes when ventricle contracts (**1**); prevents backflow (**1**)

(c) has to pump harder (**1**) to get blood all round body (**1**), not just to lungs (**1**)

71. Aerobic respiration

1 (a) oxygen and glucose (**2**)

(b) Diffusion is the movement of substances from high to low concentration. (**1**)

2 (a) mitochondria (**1**)

(b) Respiration is an exothermic process (**1**) and transfers energy by heating (**1**).

(c) One from: to build larger molecules from smaller ones (proteins from amino acids, large carbohydrates from small sugars, fats from fatty acids and glycerol) (**1**); muscle contraction (**1**); active transport (**1**)

3 (a) glucose + oxygen → carbon dioxide + water (**1**)

(b) capillaries (**1**)

4 (a) Respiration releases energy (**1**) so that metabolic processes that keep the organism alive can continue. (**1**)

(b) Two from: Plants cannot use energy from sunlight directly for metabolic processes (**1**), so they need energy from respiration for this purpose (**1**) during the day as well as at night. (**1**)

72. Anaerobic respiration

1 (a) Aerobic respiration releases more energy (**1**) per molecule of glucose (**1**).

(b) The body needs energy more quickly than aerobic respiration can supply (**1**) or if it cannot get enough oxygen to respiring cells. (**1**)

2 (a) Heart rates will increase gradually / remain low during the early laps (**1**) and increase rapidly during the final sprint (**1**) because the energy demand is low at first then increases significantly (**1**). You could also say that adrenalin might increase heart rate during the early laps.

(b) Two from: to keep heart rate relatively high (**1**) so that lactic acid is removed from muscles (**1**); because oxygen is needed to release energy needed to get rid of lactic acid (**1**)

3 (a) Three from: oxygen consumption increases during exercise (**1**) but reaches a maximum value (**1**); no more oxygen can be delivered for aerobic respiration (**1**); increased energy needed comes from anaerobic respiration (**1**)

(b) During exercise there is an increase in the concentration of lactic acid (**1**); after exercise, extra oxygen is needed to break down lactic acid. (**1**)

73. Rate of respiration

1 (a) maintains a constant temperature (**1**); because temperature can affect enzymes / change the rate of reaction (**1**)

(b) absorbs carbon dioxide produced by the seeds (**1**) so that this doesn't interfere with the movement of the blob of water (**1**)

(c) allows the pressure to be released between experiments (**1**); so the blob of water is pushed back to the start position (**1**)

2 (a) Movement of the blob of water indicates uptake of oxygen for use in respiration (**1**), so measuring the movement of the blob at intervals (**1**) allows the rate of respiration to be calculated by dividing distance moved by time taken. (**1**)

(b) Use a water bath at a range of temperatures (**1**); measure distance moved by the blob over a particular time (**1**); repeat several times at each temperature (**1**).

74. Changes in heart rate

1 (a) Stroke volume is the volume of blood pumped from the heart in one beat (**1**).

(b) (i) cardiac output = stroke volume × heart rate = 75×60 (**1**) = 4500 (**1**) cm^3 / minute (**1**)

(ii) Cardiac output increases (**1**); then two from: cells need to respire faster / need more oxygen and glucose (**1**); increased stroke volume / more blood needed for respiring cells (**1**); so heart rate must increase (**1**).

2 (a) $100 - 80 = 20$ (**1**); $20 / 80 \times 100 = 25\%$ (**1**)

(b) highest demand for oxygen / glucose / respiration (**1**)

(c) rearrange the equation to give stroke volume = cardiac output / heart rate (**1**) = 4000 / 50 = 80 (**1**) cm^3

75. Extended response – Exchange

*Answer could include the following points: (**6**)
Outline of route:
- vena cava → right atrium → right ventricle → pulmonary artery → (capillaries in) lungs → pulmonary vein → left atrium → left ventricle → aorta → rest of the body / capillaries in the body → vena cava

Answer might also include:
- valves in heart / veins prevent backflow of blood
- deoxygenated blood enters / leaves right side
- oxygenated blood enters / leaves left side
- walls of left side of heart are thicker than right side

76. Ecosystems and abiotic factors

1 **1 mark** for each

Term	Definition
Community	A single living individual
Organism	All the living organisms and the non-living components in an area
Population	All the populations in an area
Ecosystem	All the organisms of the same species in an area

2 (a) south side = 323.5 (**1**); north side = 227.6 (**1**)

(b) She is correct (**1**) because the mean percentage cover is 39% on the south side compared with 4% on the north side / 10 times greater. (**1**)

(c) temperature (**1**) because it affects enzymes / rate of reactions (**1**); OR humidity (**1**) because water required for photosynthesis / other cell processes (**1**)

77. Biotic factors

1 (a) the living parts of an ecosystem (**1**)

(b) Two from: so that they can become the new alpha male (**1**); to gain fighting skills (**1**); to become stronger (**1**)

(c) Food can often be scarce in their habitat (**1**), so large groups need to split into different areas in order to find enough food. (**1**)

2 The peacock has large, attractive tail feathers (**1**); it competes with other males for mates (**1**), so large showy tails are more attractive to female peahens (**1**). You could also suggest that the large tail feathers can be used to help scare away other male peacocks who may compete for females.

3 (a) The trees emerge through the canopy to get more light (**1**) for more photosynthesis. (**1**)

(b) The trees have deep / extensive roots (**1**) to collect minerals (**1**).

78. Parasitism and mutualism

1 Both involve interdependence / the survival of one species is closely linked with another species (**1**); in mutualism both species benefit / win-win (**1**), but in parasitism one species is harmed / one species benefits at the expense of the other / win–lose. (**1**)

2 (a) Fleas: obtain nutrients by sucking blood of the host (**1**); Animals: are harmed because they lose blood / nutrients / can catch disease from fleas (**1**)

(b) Parasitic (**1**) because fleas benefit and animals are harmed. (**1**)

3 Cleaner fish get food by eating parasites from the skin of sharks (**1**). This helps the shark because it reduces the risk of the shark being harmed by the parasites. (**1**)

4 scabies mite lives in the host and causes it harm (**1**); no benefit to the host (**1**)

79. Fieldwork techniques

1 (a) He could use a 1 m × 1 m quadrat (**1**), which he would throw at different places around the flower bed. (**1**)

(b) Using the same area means that his experiment is a fair test. (**1**)

(c) He could look at more than one area each day (**1**) and take an average number of slugs. (**1**)

2 (a) Find the total number of plants counted and the number of quadrats (**1**) and calculate a mean number of clover plants. (**1**)

(b) total size of field = 100 × 65 = 6500 m² (**1**), so number of clover plants = 6500 × 7 (**1**) = 45 500 (**1**)

3 Place quadrats at regular intervals along the transect (**1**) and measure the percentage cover of broad-leaved plants in each quadrat (**1**); record a named abiotic factor (light intensity / temperature) at each quadrat position (**1**).

80. Organisms and their environment

1 (a) Draw a line from the sea shore up the beach (at right angles to the sea line) (**1**); place quadrat at regular intervals along the line (**1**); count the limpets in each quadrat area. (**1**)

(b) (10 + 8 + 9) / 3 (**1**) = 9 (**1**)

(c) The number of limpets goes down as you travel further from the sea (**1**); this decrease is linear with the distance / it drops by 4 limpets for every 0.5 m distance travelled (**1**); limpets are more likely to survive if they live nearer the seashore. (**1**)

2 (a) Instead of placing a quadrat every 2 m, the scientist could place a quadrat every 0.5 m / 1 m / smaller distance (**1**) and use a smaller (**1**) quadrat than before.

(b) Two from: the number of bluebells increases to a maximum around 8 m into the wood (**1**); less light is available for photosynthesis (**1**) and fewer nutrients / water available deeper in the wood where there are more trees. (**1**)

81. Human effects on ecosystems

1 Advantage – reduces fishing of wild fish (**1**).

Disadvantage – one of: the waste can pollute the local area changing conditions so that some local species die out (**1**); diseases from the farmed fish (e.g. lice) can spread to wild fish and kill them (**1**)

2 Advantage – one of: may provide food for native species (**1**); may increase biodiversity (**1**)

Disadvantage – one of: may reproduce rapidly as they have no natural predators in the new area (**1**); may out-compete native species for food or other resources (**1**)

3 (a) 145 – 15 = 130; (130 / 15) × 100 (**1**) = 867% (**1**)

(b) increasing population (**1**); more food / crops needed (**1**)

(c) Excess fertiliser can be leached / washed into rivers / lakes (**1**), causing eutrophication. (**1**)

82. Biodiversity

1 (a) replanting forests where they have been destroyed (**1**)

(b) Two from: restores habitat for endangered species (**1**); reduces carbon dioxide concentration in the air (as trees photosynthesise) (**1**); reduces the effects of soil erosion (**1**); reduces range of temperature variation (**1**)

2 Some species are valuable to humans (**1**) because they are a source of new drugs / are wild varieties of crop plants / source of genes. (**1**)

3 The numbers of trees will increase because there are fewer deer to eat them (**1**). This means there will be more food for birds / bears / rabbits / insects (**1**). There will be more rabbits because there are fewer coyotes to kill / eat them (**1**). If there are more rabbits, there will be more food for coyotes / hawks / predators (**1**). More trees also mean that there will be more habitats for birds (**1**) and less soil erosion. (**1**)

83. The carbon cycle

1 (a) photosynthesis (**1**)

(b) respiration (**1**)

(c) combustion (**1**)

(d) decomposition (**1**)

2 Microorganisms are decomposers (**1**); complex carbon-containing molecules release carbon dioxide into the atmosphere (from respiration). (**1**)

3 (a) Fish carry out respiration (**1**); respiration releases carbon dioxide into the water (**1**); plants absorb the carbon dioxide (**1**), which is used in photosynthesis. (**1**)

(b) Three from: if there are not enough fish / snails / aquatic animals in the tank there will not be enough carbon dioxide (**1**); so there is less photosynthesis by plants (**1**); so less oxygen is released for fish / snails / aquatic animals (**1**); less food for animals as fewer plants (**1**); plants and fish / snails / aquatic animals die (**1**)

84. The water cycle

1 (a) Three from: evaporation from land (**1**) / sea (**1**); transpiration from plants (**1**) / animal sweat (**1**) / animal breath (**1**)

(b) Water vapour condenses to form clouds (**1**); water cools to form precipitation / rain / snow (**1**) that returns the water to Earth. (**1**)

2 A lot of water evaporates from a golf course so this will lead to more water in the atmosphere (**1**); water levels fall in the river as water is removed for watering the golf course (**1**), so animals or plants living in the river might die. (**1**)

3 Advantage: sea water is made potable / safe to drink (**1**)

Disadvantage: needs a lot of energy / fuel / it is expensive (**1**)

85. The nitrogen cycle

1 C (**1**)

2 (a) Nitrogen fixation by soil bacteria (**1**)

(b) Three from: nitrates are absorbed by roots (**1**) by active transport (**1**); plants need nitrogen for making amino acids / proteins (**1**); but can only take in nitrogen in the form of nitrate / ammonium (ions / salts) (**1**)

(c) Amount of nitrate in soil is reduced (**1**) because bacteria convert nitrates in the soil into nitrogen gas in the air (**1**).

3 (a) Three from: Plants such as clover have nitrogen-fixing bacteria in their roots (**1**), so they can be grown and ploughed back into the soil (**1**) where they are decomposed (**1**) to add nitrates. (**1**)

86. Extended response – Ecosystems and material cycles

* Answer could include the following points: **(6)**

- Fish farming can reduce biodiversity by introducing just one species.
- Waste and diseases can affect wild populations.
- Introduction of non-native species might lead to competition with native species.
- Reduces fishing of wild fish.
- Fertilisers can cause eutrophication leading to loss of biodiversity in nearby water.
- You could also talk about conservation, reforestation, captive breeding.

Chemistry

87. Formulae

1 C **(1)**

2 An element is a substance made from atoms **(1)** with the same number of protons. **(1)**

3 (a) Compounds contain two or more different elements chemically joined together **(1)**, but Cl_2 shows that chlorine contains only (two atoms of) one element. **(1)**
   (b) Cl_2 shows that chlorine exists as two chlorine atoms joined together **(1)**, and chlorine is a non-metal (and nearly all non-metal elements are molecular). **(1)**

4 **1 mark** for each correct formula: water, H_2O; carbon dioxide, CO_2; methane, CH_4; sulfuric acid, H_2SO_4; sodium, Na

5 (a) 3 **(1)**   (b) 7 **(1)**

6 It contains one carbon atom and three oxygen atoms **(1)**; it has two negative charges overall. **(1)**

88. Equations

1 B **(1)**

2 (a) sodium hydroxide + hydrochloric acid → sodium chloride + water **(1)**
   (b) $NaOH + HCl → NaCl + H_2O$ **(1)**

3 (a) **1 mark** for all four state symbols in the correct order: (s), (l), (aq), (g)
   (b) There are the same numbers of atoms **(1)** of each element on both sides of the equation. **(1)**

4 (a) $2Cu + O_2 → 2CuO$ **(1)**
   (b) $2Al + Fe_2O_3 → Al_2O_3 + 2Fe$ **(1)**
   (c) $Mg + 2HNO_3 → Mg(NO_3)_2 + H_2$ **(1)**
   (d) $Na_2CO_3 + 2HCl → 2NaCl + H_2O + CO_2$ **(1)**
   (e) $4Fe + 3O_2 → 2Fe_2O_3$ **or** $Fe + 1½O_2 → Fe_2O_3$ **(1)**
   (f) $Cl_2 + 2NaBr → 2NaCl + Br_2$ **(1)**

89. Ionic equations

1 An ion is a charged particle **(1)** formed when an atom or group of atoms loses or gains electrons. **(1)**

2 (a) silver ion Ag^+ **(1)**; iodide ion I^- **(1)**
   (b) $Ag^+(aq) + I^-(aq) → AgI(s)$
   1 mark for balanced equation, **1 mark** for state symbols

3 (a) hydrogen ion H^+ **(1)**; carbonate ion CO_3^{2-} **(1)**
   (b) $2H^+ + CO_3^{2-} → H_2O + CO_2$
   1 mark for formulae, **1 mark** for balancing

4 (a) $Fe^{2+} + 2OH^- → Fe(OH)_2$ **(1)**
   (b) $Fe^{3+} + 3OH^- → Fe(OH)_3$ **(1)**

5 (a) $H^+ + OH^- → H_2O$ **(1)**
   (b) water **(1)**

6 (a) K^+ **(1)**, Br^- **(1)**, Cl^- **(1)**
   (b) $Cl_2 + 2Br^- → 2Cl^- + Br_2$
   1 mark for Cl_2 and Br_2, **1 mark** for balancing

90. Hazards, risk and precautions

1 A hazard is something that could cause damage to / harm someone / something **(1)** or cause adverse health effects. **(1)**

2 the chance that someone or something will be harmed **(1)**, if exposed to a hazard **(1)**

3 to indicate the dangers associated with the contents **(1)**; to inform people about safe-working precautions with these substances **(1)**

4 **1 mark** for each correct line (if more than four new lines drawn, subtract **1 mark** for each extra line):

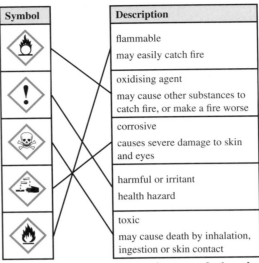

Symbol	Description
	flammable may easily catch fire
	oxidising agent may cause other substances to catch fire, or make a fire worse
	corrosive causes severe damage to skin and eyes
	harmful or irritant health hazard
	toxic may cause death by inhalation, ingestion or skin contact

5 precaution for **1 mark** with appropriate reason for **1 mark**, e.g. wear gloves to avoid skin contact because concentrated nitric acid is corrosive / nitrogen dioxide is toxic; work in a fume cupboard to avoid breathing in nitrogen dioxide (which is toxic)

91. Atomic structure

1 A **(1)**

2 **1 mark** for each correct row

Particle	proton	neutron	electron
Relative mass	1	1	1 / 1836
Relative charge	+1	0	−1
Position	nucleus	nucleus	shell

3 The atom contains equal numbers of protons and electrons / 1 proton and 1 electron **(1)**; protons and electrons have equal but opposite charges / relative charge of proton is +1 and relative charge of electron is −1. **(1)**

4 These particles were not discovered until later. **(1)**

5 $(2.70 × 10^{-10}\ m) / (1.03 × 10^{-14}\ m) = 26\,200$
   1 mark for working out, **1 mark** for correct answer to 3 significant figures

6 (a) Most of an atom is empty space. **(1)**
   (b) The nucleus is positively charged **(1)** because like charges repel. **(1)**
   (c) Only a few particles come close to the nucleus **(1)** because it is very small compared with the rest of the atom. **(1)**

92. Isotopes

1 The mass number of an atom is the total number of protons and neutrons (in the nucleus). **(1)**

2 A **(1)**

3 a substance whose atoms all have the same number of protons **(1)** which is unique to that element / different from all other elements **(1)**

4 (a) **1 mark** for each correct row:

Isotope	Protons	Neutrons	Electrons
hydrogen-1	1	0	1
hydrogen-2	1	1	1
hydrogen-3	1	2	1

   (b) Their nuclei have the same number of protons **(1)** but different numbers of neutrons. **(1)**

5 Some elements have different isotopes **(1)** so their relative atomic masses are a (weighted) mean value. **(1)**

6 Relative abundance of neon-22 = 9.5% **(1)**
   Mass of 100 atoms = $(20 × 90.5) + (22 × 9.5) = 2019$ **(1)**

A_r = 2019 / 100 = 20.19
= 20.2 (to one decimal place) **(1)**

93. Mendeleev's table

1 (a) D **(1)**
   (b) the properties of the elements and their compounds **(1)**

2 (a) Two similarities for **1 mark** each, e.g. elements put into groups; elements put into periods; elements with similar properties put into the same groups
   (b) Three differences for **1 mark** each, e.g. Mendeleev's table had gaps / fewer elements / did not include the noble gases; was arranged in order of increasing relative atomic mass (not atomic number); did not have a block of transition metals; had two elements in some spaces.

3 (a) The relative atomic mass of tellurium is greater than that of iodine **(1)** and Mendeleev thought that he had arranged the elements in order of increasing relative atomic mass. **(1)**
   (b) Atoms of tellurium have fewer protons / lower atomic number than iodine / tellurium has some isotopes with higher masses than iodine **(1)**; elements are arranged in order of increasing atomic number in the modern periodic table. **(1)**

4 He predicted properties of undiscovered elements. **(1)**

94. The periodic table

1 B **(1)**

2 (a) **A** and **B (1)**
   (b) **A, B, C, D** (all four for **1 mark**)
   (c) **B** and **E (1)**

3 (a) the position of an element **(1)** on the periodic table **(1)**
   (b) the number of protons **(1)** in the nucleus of an atom **(1)**
   (c) discovery of protons / discovery of atomic structure / better technology available to study atoms **(1)**

4 Each element has a unique atomic number / unique number of protons / unique space on the periodic table **(1)**; there are only six spaces (between sodium and argon) so there can be only six elements. **(1)**

95. Electronic configurations

1 (a) 2.1 **(1)**
   (b) There are three electrons, so there must be three protons **(1)**, so the four shaded circles must be neutrons. **(1)**
   (c) two shells **(1)**; two electrons in first shell, six in the second **(1)**, e.g.

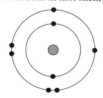

2 (a) Both have 7 electrons **(1)** in their outer shell. **(1)**
   (b) Number of occupied shells is the same as the period number **(1)**; fluorine has two occupied shells and chlorine has three. **(1)**

3 (a) 2.8.8.2 **(1)**
   (b) 2.8.5 **(1)**

4 group 0 / 8 / 18 **(1)** because it has a full outer shell **(1)**

96. Ions

1 D **(1)**

2 (a) 10 **(1)**
   (b) 2.8 **(1)**

3 **1 mark** for each correct row:

Ion	Atomic number	Mass number	Protons	Neutrons	Electrons
N^{3-}	7	15	7	8	10
K^+	19	40	19	21	18
Ca^{2+}	20	40	20	20	18
S^{2-}	16	32	16	16	18
Br^-	35	81	35	46	36

4 three shells and 2.8.7 electrons in the chlorine atom **(1)**; three shells and 2.8.8 electrons in the chloride ion **(1)**; brackets with negative sign **(1)**, e.g.

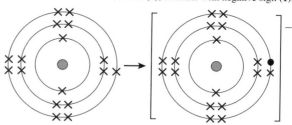

97. Formulae of ionic compounds

1 D **(1)**

2 **1 mark** for each correct formula:

	Cl^-	S^{2-}	OH^-
K^+	KCl	K_2S	KOH
Ca^{2+}	$CaCl_2$	CaS	$Ca(OH)_2$
Fe^{3+}	$FeCl_3$	Fe_2S_3	$Fe(OH)_3$
NH_4^+	NH_4Cl	$(NH_4)_2S$	NH_4OH

	NO_3^-	SO_4^{2-}
K^+	KNO_3	K_2SO_4
Ca^{2+}	$Ca(NO_3)_2$	$CaSO_4$
Fe^{3+}	$Fe(NO_3)_3$	$Fe_2(SO_4)_3$
NH_4^+	NH_4NO_3	$(NH_4)_2SO_4$

3 (a) $2Mg + O_2 \rightarrow 2MgO$ (**1 mark** for correct formulae, **1 mark** for balancing)
   (b) (i) The nitrogen atom has five electrons in its outer shell **(1)**; it gains three electrons to complete its outer shell / form an ion. **(1)**
       (ii) Mg_3N_2 **(1)**
       (iii) Nitride ion contains only nitrogen **(1)** but the nitrate ion also contains oxygen. **(1)**

4 **1 mark** for each correct name:

	S^{2-}	SO_4^{2-}	Cl^-	ClO_3^-
Name	sulfide	sulfate	chloride	chlorate

98. Properties of ionic compounds

1 A **(1)**

2 (a) + and − signs drawn as shown **(1)**

   (b) strong electrostatic forces **(1)** between oppositely charged ions **(1)**

3 (a) strong (electrostatic) forces of attraction / ionic bonds between **ions** **(1)** which need a lot of heat / energy to break / overcome **(1)**
   (b) MgO has stronger (ionic) bonds than NaCl. **(1)**

4 (a) When calcium chloride is a liquid, its ions are free to move **(1)** and ions are charged particles. **(1)**
   (b) Its **ions** are not free to move / are held in fixed positions / a lattice. **(1)**
   (c) Dissolve it in water. **(1)**

99. Covalent bonds

1 B **(1)**

2 when a **pair** of electrons **(1)** is **shared** between two atoms **(1)**

3 (a) A fluorine atom has one unpaired electron in its outer shell **(1)** so it shares this electron with another fluorine atom. **(1)**
   (b) There is one covalent bond **(1)** between two hydrogen atoms in a molecule. **(1)**
   (c) (i) correct diagram, e.g.

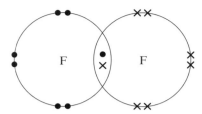

one pair of dots and crosses in shared area (**1**); three pairs of dots and three pairs of crosses (**1**)

(ii) correct diagram, e.g.

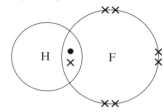

one pair of dots and crosses in shared area (**1**); (three) pairs of dots / crosses on F only (**1**)

4 (a) correct diagram, e.g.

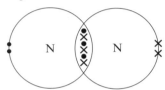

three pairs of dots and crosses in shared area (**1**); one pair of dots and one pair of crosses (**1**)

(b) N≡N (**1**)

100. Simple molecular substances

1 C (**1**)

2 substance B because: it has a low melting point / lowest melting point (**1**); it is (almost) insoluble in water (**1**); it does not conduct electricity when solid or liquid (**1**)

3 (a) Sulfur hexafluoride molecules are not charged (**1**) and have no electrons that are free to move. (**1**)

(b) The intermolecular forces between water and sulfur hexafluoride molecules (**1**) are weaker than those between water molecules (**1**) and those between sulfur hexafluoride molecules. (**1**)

4 As the relative formula mass increases so does the boiling point (**1**) because the intermolecular forces become stronger / there are more intermolecular forces to overcome. (**1**)

101. Giant molecular substances

1 B (**1**)

2 (a) carbon (**1**)

(b) four (**1**)

(c) giant molecular / giant covalent (**1**)

3 (a) Each carbon atom forms four covalent bonds (**1**); in a regular lattice structure (**1**); the bonds are strong / it is difficult to separate atoms from the structure. (**1**)

(b) The layers in graphite can slide over each other (**1**) because there are weak forces between the layers. (**1**)

(c) Each carbon atom only forms three bonds (**1**) so there are (outer) electrons that are delocalised / free to move. (**1**)

102. Other large molecules

1 C (**1**)

2 (a) carbon (**1**)

(b) covalent (**1**)

3 (a) Graphene has covalent (**1**) bonds in a giant / lattice structure (**1**), and these bonds are strong. (**1**)

(b) Each carbon atom only forms three bonds (**1**) so there are (outer) electrons that are delocalised / free to move. (**1**)

4 It has a simple molecular structure (**1**) with intermolecular forces (**1**) that are weak / easily overcome. (**1**)

103. Metals

1 **1 mark** for each correct tick (deduct **1 mark** for each tick over four ticks)

	Low melting points	High melting points	Good conductors of electricity	Poor conductors of electricity
Metals		✓	✓	
Non-metals	✓			✓

2 having a high mass (**1**) for its volume / size (**1**) OR small volume (**1**) for its mass (**1**)

3 (a) + / 2+ inside the circles (**1**); delocalised electrons / sea of electrons (**1**)

(b) Layers of atoms / (positive) ions (**1**) can slide over each other. (**1**)

(c) Delocalised electrons (**1**) can move through the structure. (**1**)

4 Metallic bonds require a lot of energy to break (**1**) because there are strong electrostatic forces of attraction (**1**) between positive metal ions and delocalised electrons. (**1**)

5 Calcium reacts with water (**1**) to form a soluble product / soluble calcium hydroxide. (**1**)

104. Limitations of models

1 B (**1**)

2 (a) A, B, C (**1**)   (d) B (**1**)

(b) A, B (**1**)   (e) C, D (**1**)

(c) C, D (**1**)

3 Answer must include at least one advantage and one disadvantage to gain full marks. (**3**)

Advantages of ball and stick include, for **1 mark** each: shape of molecule shown; can be modelled in three dimensions (e.g. using a modelling kit).

Disadvantages of ball and stick include, for **1 mark** each: element symbols not shown; bonding electrons not shown; non-bonding electrons not shown.

105. Relative formula mass

1 (a) 18 (**1**)   2 (a) 74 (**1**)

(b) 64 (**1**)   (b) 78 (**1**)

(c) 102 (**1**)   (c) 164 (**1**)

(d) 53.5 (**1**)   (d) 132 (**1**)

(e) 111 (**1**)   (e) 342 (**1**)

(f) 133.5 (**1**)

106. Empirical formulae

1 (a) to make sure the reaction had finished / all the magnesium had reacted (**1**)

(b) use tongs / allow the crucible to cool down (**1**) to prevent skin burns (**1**)

(c) mass of magnesium used = 0.25 g (**1**)

mass of oxygen reacted = 0.16 g (**1**)

Mg	**O**
0.25 / 24 = 0.0104	0.16 / 16 = 0.010 (**1**)
0.0104 / 0.010 = 1.04 (**1**)	0.010 / 0.010 = 1.00 (**1**)

Empirical formula is MgO (**1**)

2 correct answer without working for one mark only, otherwise:

Fe	**Cl**
11.2 / 56 = 0.20	21.3 / 35.5 = 0.60 (**1**)
0.20 / 0.20 = 1.0	0.60 / 0.20 = 3.0 (**1**)

Empirical formula is $FeCl_3$ (**1**)

3 M_r of NO_2 = 46 (**1**)

Ratio is 92 / 46 = 2, so molecular formula is N_2O_4 (**1**)

107. Conservation of mass

1 (a) It is a closed system because no substances can enter or leave. (**1**)

(b) The total mass will stay the same (**1**) because mass is conserved in chemical reactions. (**1**)

2 35.1 g **(1)** **First mark is for the working out, shown in the Guided example.**

3 M_r of $O_2 = 32$ and M_r of $MgO = 40$ **(1)**

   $(1 \times 32) = 32$ g of O_2 makes $(2 \times 40) = 80$ g of MgO **(1)**

   12.6 g of O_2 makes $80 \times (12.6 / 32) = 31.5$ g of MgO **(1)**

4 M_r of $CaCO_3 = 100$ and M_r of $CaO = 56$ **(1)**

   $(1 \times 100) = 100$ kg of $CaCO_3$ makes $(1 \times 56) = 56$ kg of CaO **(1)**

   12.5 kg of $CaCO_3$ makes $56 \times (12.5 / 100) = 7.00$ kg of CaO **(1)**

108. Reacting mass calculations

1 D **(1)**

2 (a) The mass of copper oxide formed increases as the mass of copper carbonate increases. **(1)**
   but
   The mass of copper oxide formed is **directly** proportional to the mass of copper carbonate. **(2)**

   (b) Copper carbonate is the limiting reagent / only reagent **(1)** so atoms can come only from copper carbonate. **(1)**

3 mass of chlorine reacted $= 32.5$ g $- 11.2$ g
   $= 21.3$ g **(1)**

   all three amounts for **1 mark**:

   amount of iron $= 11.2 / 56 = 0.2$ mol

   amount of chlorine $= 21.3 / 71 = 0.3$ mol

   amount of iron(III) chloride $= 32.5 / 162.5 = 0.2$ mol

   ratio of $Fe : Cl_2 : FeCl_3 = 0.2 : 0.3 : 0.2$ **(1)**

   whole number ratio of $Fe : Cl_2 : FeCl_3 = 2 : 3 : 2$

   so equation must be: $2Fe + 3Cl_2 \rightarrow 2FeCl_3$ **(1)**

109. Concentration of solution

1 (a) 2 dm^3 **(1)**
   (b) 0.5 dm^3 **(1)**
   (c) 0.025 dm^3 **(1)**

2 concentration $= (10 / 250) \times 1000$
   $= 40$ g dm^{-3} **(1)**

3 (a) 50 g dm^{-3} **(1)**
   (b) 36.5 g dm^{-3} **(1)**
   (c) 10 g dm^{-3} **(1)**

4 (a) mixture of a solute and water / solution in which the solvent is water **(1)**
   (b) 100 g **(1)**
   (c) mass of NaOH in 50 cm$^3 = 40 \times 50 / 1000 = 2$ g **(1)**

   new concentration $=$ (mass / 250) $\times 1000$
   $= (2 / 250) \times 1000 = 8$ g dm^{-3} **(1)**

110. Avogadro's constant and moles

1 Avogadro's number / Avogadro's constant **(1)**

2 (a) 65.5 g **(1)**
   (b) 48 g **(1)**
   (c) 5.4 g **(1)**

4 (a) 3.0 mol **(1)**
   (b) 4.0 mol **(1)**
   (c) 0.600 mol **(1)**

3 (a) 0.5 mol **(1)**
   (b) 2.5 mol **(1)**
   (c) 0.25 mol **(1)**

5 1.35×10^{24} **(1)**

6 15 mol **(1)**

111. Extended response – Types of substance

*Answer could include the following points: **(6)**

Graphite uses and properties:

- lubricant because it is soft / the softest in the table / 10 times softer than copper
- electrodes because it is a good conductor of electricity / conductivity is 100 times less than copper / 100 million times better than diamond

Graphite bonding and structure:

- giant covalent / giant molecular structure
- strong covalent bonds
- each carbon atom is bonded to three other carbon atoms
- layers of carbon atoms
- weak forces between layers

- layers can slide past each other (making it slippery so that it can be used as a lubricant)
- one free electron per carbon atom
- delocalised electrons
- electrons can move through the structure (allowing it to conduct electricity for use as an electrode)

Diamond uses and properties:

- cutting tools because it is very hard / 100 times harder than copper / 1000 times harder than graphite

Diamond bonding and structure:

- giant covalent / giant molecular structure
- strong covalent bonds
- each carbon atom is bonded to four other carbon atoms
- three-dimensional lattice structure
- a lot of energy is needed to break the many strong bonds
- rigid structure

112. States of matter

1 C **(1)**

2 (a) freezing / solidifying **(1)**
   (b) condensing **(1)**

3 The chemical composition is unchanged. **(1)**

4 (a) solid: close together and regular pattern **(1)**; vibrate about fixed positions **(1)**

   liquid: close together and random **(1)**; move around each other / slide about in groups **(1)**

   gas: far apart and random **(1)**; move quickly in all directions **(1)**

   (b) gas **(1)** because the particles are moving freely / moving fastest / have the most kinetic energy **(1)**

5 The arrangement changes from random to regular **(1)** and the movement changes from moving around each other (in groups) to vibrating about fixed positions. **(1)**

6 liquid **(1)**

113. Pure substances and mixtures

1 (a) The atoms of an element all have the same atomic number / number of protons **(1)** but atoms of Na and Cl_2 have different atomic numbers / numbers of protons. **(1)**
   (b) substance containing two or more elements **(1)** **chemically** combined / joined together **(1)**

2 (a) **B** 0.24 and **C** 0.03 **(1)**
   (b) None of the samples is pure **(1)**; all contain some residue / dissolved solid / solid mixed in. **(1)**

3 Pure substances (e.g. tin and silver) have a sharp melting point **(1)** but mixtures (e.g. solder) melt over a range of temperatures. **(1)**

4 (a) The salts increase the boiling point. **(1)**
   (b) The boiling point will increase further **(1)** because the concentration of dissolved salt will increase **(1)** as the water leaves the seawater. **(1)**

114. Distillation

1 C **(1)**

2 (a) The vapour is cooled **(1)** and condensed / turned from gas to liquid. **(1)**
   (b) The temperature of the water increases **(1)** because it is heated up by the vapour / energy is transferred from the vapour to the water by heating. **(1)**

3 (a) They have different boiling points. **(1)**
   (b) Ethanol, because it has a lower boiling point than water **(1)**, it boils / evaporates first. **(1)**
   (c) One of the following for **1 mark**:
     - more energy is transferred by heating
     - hot vapour and cold water flow in opposite directions
     - the condenser will be full of cold water / will not contain any air

115. Filtration and crystallisation

1 second box ticked **(1)**; fourth box ticked **(1)** **Deduct one mark for each extra tick above two ticks.**

2 (a) $2KI(aq) + Pb(NO_3)_2(aq) \rightarrow 2KNO_3(aq) + PbI_2(s)$

265

1 mark for correct balancing; **1 mark** for correct state symbols

    (b) (i) Its particles are too large to pass through. **(1)**

       (ii) to remove excess potassium iodide solution / lead nitrate solution / potassium nitrate solution **(1)**

3 (a) filtration / filtering **(1)**

    (b) (i) **Water** evaporates **(1)**; solution becomes saturated **(1)**; crystals form as more **water** evaporates **(1)**.

       (ii) to dry the crystals **(1)**

    (c) Two from the following, for **1 mark** each: heat the solution slowly; do not evaporate all of the water; leave to cool down so crystals form

116. Paper chromatography

1 (a) Pencil does not dissolve in the solvent. **(1)**

    (b) mixture because it contains more than one substance / four substances **(1)**; pure substances only contain one substance **(1)**

    (c) **A** and **B** **(1)**

    (d) The orange squash does contain X because it contains spots with the same R_f values / which move the same distances **(1)** as the two spots in X. **(1)**

    (e) It is insoluble in the solvent / mobile phase / has very strong bonds with the stationary phase / has very weak bonds with the mobile phase. **(1)**

2 correctly measured distance from start line to spot **(1)**

    correctly measured distance from start line to solvent front **(1)**

    $R_f = 0.8$ (no units) **(1)**

    Actual distances measured will depend on print scale, but final R_f should equal 0.8.

117. Investigating inks

1 (a) distance travelled by the spot / dye **(1)**; distance travelled by the solvent / solvent front **(1)**

    (b) The measurements will be more precise / have a higher resolution **(1)** so the R_f value will be more accurate / closer to the true value. **(1)**

2 (a) to stop the solution boiling over (into the condenser) **(1)** so that the solvent collected is not contaminated with the solution / so that the vapour is not produced faster than it can be condensed **(1)**

    (b) The apparatus will get very hot / solvent vapour (e.g. steam) will escape. **(1)**

3 (a) (highly) flammable **(1)**

    (b) Two from the following, for **1 mark** each (precaution and reason needed for each mark):

- avoid naked flames because the liquid is flammable
- wear gloves because it can cause skin dryness
- work in a fume cupboard / keep lab well ventilated because vapour causes dizziness

118. Drinking water

1 A **(1)**

2 (a) to sterilise the water / to kill microbes **(1)**

    (b) The concentration of chlorine is very low **(1)**; it is low enough to kill microbes without being harmful to people. **(1)**

3 sedimentation **(1)** to remove large insoluble particles **(1)**

    filtration **(1)** to remove small insoluble particles **(1)**

4 Unlike tap water, distilled water does not contain dissolved salts **(1)**. These would interfere with the analysis / react with test substances / give a false-positive result. **(1)**

5 (a) **simple** distillation **(1)**

    (b) A lot of energy is needed / a lot of fuel is needed. **(1)**

6 $Al_2(SO_4)_3(aq) + 6H_2O(l) \rightarrow 2Al(OH)_3(s) + 3H_2SO_4(aq)$ **(1)**

119. Extended response – Separating mixtures

Answer could include the following points: **(6)**

Physical states:

- substance A is solid; substances B and C are liquids

Separating A from B and C:

- substance A is insoluble in B and C
- so it cannot be separated by paper chromatography

- but it can be separated from B and C by filtration
- substance A collects as a residue in the filter paper
- it can be washed with B or C on the filter paper
- then dried in a warm oven
- below 115 °C to stop it melting

Separating B and C:

- (after filtration) the filtrate is a mixture of substances B and C
- they have different boiling points
- so they can be separated by **fractional** distillation
- substance B has the lower boiling point
- substance B distils off first (and can be collected)
- continue heating to leave substance C in the flask

120. Acids and alkalis

1 (a) The green colour means that the indicator is neutral **(1)**; neutral pH is 7. **(1)**

    (b) red / orange **(1)**

    (c) hydrogen ion **(1)**

2 (a) $2Mg + O_2 \rightarrow 2MgO$

    1 mark for correct formulae, **1 mark** for correct balancing

    (b) It is alkaline / contains an alkali / has a pH greater than 7. **(1)**

3 $NaOH(aq) \rightarrow Na^+(aq) + OH^-(aq)$ **(1)**

4 litmus: blue, red **(1)**; methyl orange: yellow, red **(1)**, phenolphthalein: pink, colourless **(1)**

5 (a) It decreases. **(1)**

    (b) It increases. **(1)**

121. Strong and weak acids

1 D **(1)**

2 Nitric acid is fully / completely dissociated into ions in aqueous solution. **(1)**

3 (a) The reaction is reversible. **(1)**

    (b) It is partially dissociated into ions (in aqueous solution). **(1)**

4 The concentrated solution contains a greater amount of dissolved sodium hydroxide **(1)** in the same volume. **(1)**

5 (a) At the same concentration / at 0.20 mol dm^{-3} **(1)** the pH of hydrochloric acid is lower than the pH of ethanoic acid. **(1)**

    (b) Its pH increases by 1 **(1)** when the acid is diluted by a factor of 10 / 10 times. **(1)**

    The pH increases as the acid is diluted, for **1 mark** only.

    (c) The concentration of hydrogen ions / H$^+$ ions is the same. **(1)**

122. Bases and alkalis

1 B **(1)**

2 (a) $\mathbf{Na_2CO_3 + 2HNO_3 \rightarrow 2NaNO_3 + CO_2 + H_2O}$

    1 mark for correct formulae, **1 mark** for balancing

    (b) bubbles **(1)**; powder disappears / dissolves **(1)**

    (c) (bubble the gas through) limewater **(1)** which turns milky / cloudy white **(1)**

3 (a) calcium chloride solution **(1)**

    (b) (i) hydrogen **(1)**

       (ii) **lighted** splint (ignites the gas) **(1)** with a (squeaky) pop **(1)**

4 (a) any substance that reacts with an acid **(1)** to form a salt and water only **(1)**

    (b) alkali **(1)**

    (c) zinc sulfate **(1)**

123. Neutralisation

1 Hydrogen ions / H$^+$ ions from the acid **(1)** react with hydroxide ions / OH$^-$ ions from the alkali **(1)** to form water **(1)**. **H$^+$(aq) + OH$^-$(aq) $\rightarrow$ H$_2$O(l) without reference to source of ions scores 2 marks.**

2 (a) $CaO + 2HCl \rightarrow CaCl_2 + H_2O$

    1 mark for formulae, **1 mark** for balancing

    (b) $Ca(OH)_2 + 2HCl \rightarrow CaCl_2 + 2H_2O$

    1 mark for formulae, **1 mark** for balancing

3 to make sure the readings / pH values are accurate **(1)**

4   points plotted accurately (± ½ square) **(1)**; line of best fit **(1)**; axes labelled using table headings **(1)**

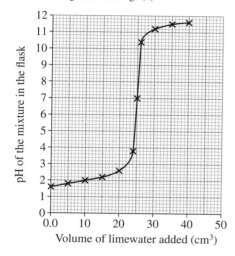

124.   Salts from insoluble bases

1   (a) to react with **all** the acid **(1)** so that only a salt and water are left (with an excess of solid) **(1)**
    (b) to make the reaction happen faster / to increase the rate of reaction **(1)**
    (c) filtration / filtering **(1)**
    (d) crystallisation / evaporation **(1)**

2   (a) measuring cylinder / pipette / burette **(1)**
    (b) two improvements with reasons: **1 mark** for improvement, **1 mark** for its reason(s), e.g.

- stir, to mix the reactants
- add copper oxide one spatula at a time, to reduce waste
- warm the acid first, to make the reaction happen faster
- add copper oxide one spatula at a time until some is left over, to make sure that all the acid has reacted

    (c) Heat the evaporating basin over a hot water bath **(1)**, allow to cool, and pour away excess water / dry crystals between paper / dry crystals in a warm oven. **(1)**

125.   Salts from soluble bases

1   C **(1)**

2   top label: burette; **(1)** bottom label: (conical) flask **(1)**

3   (a) hydrochloric acid **(1)**
    (b) (volumetric) pipette **(1)**
    (c) pink to colourless **(1) Both colours needed in the correct order for the mark.**
    (d) (i)   to get an idea of how much acid must be added **(1)**
        (ii) rough: 26.10; run 1: 24.90; run 2: 24.40; run 3: 24.50 **(1)**
        (iii) (ignore 24.90); (24.40 + 24.50) / 2 = 24.45 cm^3

        1 mark for working out, **1 mark** for answer

    (e) Repeat the titration without the indicator **(1)**, add the mean titre volume of hydrochloric acid (to 25.0 cm^3 of sodium hydroxide solution). **(1)**

126.   Making insoluble salts

1   B **(1)**

2   D **(1)**

3   (a) calcium nitrate / calcium chloride **(1)** with sodium hydroxide / potassium hydroxide / ammonium hydroxide **(1)**
    (b) Answer depends on the combination used in part (**a**), e.g. sodium nitrate / potassium nitrate / ammonium nitrate (if calcium nitrate used); sodium chloride / potassium chloride / ammonium chloride (if calcium chloride used). **(1)**

4   (a) $Na_2CO_3(aq) + CaCl_2(aq) \rightarrow 2NaCl(aq) + CaCO_3(s)$

    1 mark for correct equation, **1 mark** for correct state symbols

    (b) Dissolve sodium carbonate and calcium chloride in water then mix **(1)**; filter to separate the precipitate of calcium carbonate **(1)**; wash

the precipitate with water (e.g. on the filter paper) **(1)**; then dry in a warm oven / dry between pieces of filter paper / leave to dry. **(1)**

5   Sulfuric acid contains sulfate ions **(1)** which react with lead ions **(1)** to form insoluble lead sulfate / an insoluble product. **(1)**

127.   Extended response – Making salts

*Answer could include the following points: (**6**)

The titration:

- rinse a burette with dilute hydrochloric acid, then fill the burette with the acid
- measure 25 cm^3 of sodium hydroxide solution using a pipette
- into a conical flask
- conical flask on a white tile
- add a few drops of phenolphthalein indicator / methyl orange indicator
- record the start reading on the burette
- add dilute hydrochloric acid from the burette to the sodium hydroxide solution
- swirl the flask
- add drop by drop near the end-point
- stop when colour changes / pink to colourless (phenolphthalein) / yellow to orange (methyl orange)
- record the end reading on the burette
- repeat the experiment
- until consistent / concordant results are obtained

Using the titre:

- add 25 cm^3 of sodium hydroxide to the flask
- do not add indicator
- add the titre / mean titre volume of dilute hydrochloric acid from the burette

Producing the crystals:

- pour the mixture into an evaporating basin
- heat over a hot water bath
- until most of the water has evaporated
- allow to cool and pour away excess water
- dry crystals between filter paper / in a warm oven

128.   Electrolysis

1   D **(1)**

2   An electrolyte is an ionic compound **(1)** in the molten / liquid state or dissolved in water. **(1)**

3   MnO_4^- ions / manganate(VII) ions **(1)** move to the positively charged electrode / oppositely charged electrode. **(1)**

4   cathode: zinc **(1)**; anode: bromine **(1) not bromide**

5   (a) cathode **(1)**
    (b) Sodium ions are reduced **(1)** because they gain electrons. **(1)**

6   (a) $Al^{3+} + 3e^- \rightarrow Al$ **(1)**
    (b) $2O^{2-} \rightarrow O_2 + 4e^-$ **(2)**

    1 mark for unbalanced equation

129.   Electrolysing solutions

1   (a) D **(1)**
    (b) Some water molecules dissociate / ionise. **(1)**

        $H_2O \rightleftharpoons H^+ + OH^-$ **(1)**

    (c) $2Cl^- \rightarrow Cl_2 + 2e^-$ **(2)**

        1 mark for unbalanced equation

2   (a) Na^+, Cl^- **(1)** H^+, OH^- **(1)**
    (b) (i)   chlorine **(1)**
        (ii) hydrogen **(1)**
    (c) hydroxide ions / OH^- ions are left over / in excess **(1)**; alkalis in solution are a source of hydroxide ions / OH^- ions **(1)**

3   Hydroxide ions / OH^- ions from the water **(1)** are oxidised / lose electrons. **(1)**

    equation for **2 marks**: $4OH^- \rightarrow 2H_2O + O_2 + 4e^-$

4   Water is a covalent compound / a poor conductor of electricity / contains very few mobile ions **(1)**; sulfuric acid increases the concentration of ions / increases the conductivity of the water. **(1)**

130. Investigating electrolysis

1 (a) anode (1) because oxygen is formed from negatively charged ions / hydroxide ions (1)

(b) $Cu \rightarrow Cu^{2+} + 2e^-$ (2)

   1 mark for unbalanced equation

2 (a) $Cu^{2+} + 2e^- \rightarrow Cu$ (2)

   1 mark for unbalanced equation

(b) (i) time (1)

   (ii) gain in mass by copper cathode (1)

   (iii) $(0.15 - 0.04) / (0.8 - 0.2) = 0.11 / 0.6$ (1)
   $= 0.18$ g / A

   1 mark for correct answer, **1 mark** for 2 significant figures

131. Extended response – Electrolysis

*Answer could include the following points: (6)

Copper chloride powder:

- its ions are not free to move
- in the solid state
- so there are no visible changes

Copper chloride solution:

- its ions are free to move
- when dissolved in water / in solution
- brown solid is copper
- yellow-green gas is chlorine

Electrode reactions:

- positively charged ions / copper ions attracted to negative electrode / cathode
- $Cu^{2+} + 2e^- \rightarrow Cu$
- negatively charged ions / chloride ions attracted to positive electrode / anode
- $2Cl^- \rightarrow Cl_2 + 2e^-$
- overall reaction: $CuCl_2(aq) \rightarrow Cu(s) + Cl_2(g)$

132. The reactivity series

1 (a) D (1)

(b) Two from the following, for **1 mark** each: temperature (of water / acid); mass of metal; surface area of metal; amount / moles of metal

2 (a) hydrogen (1)

(b) $Mg + 2H_2O \rightarrow Mg(OH)_2 + H_2$

   1 mark for correct formulae, **1 mark** for balancing

(c) magnesium oxide (1)

3 (a) $Al_2O_3 + 3H_2SO_4 \rightarrow Al_2(SO_4)_3 + 3H_2O$ (1)

(b) $2Al(s) + 3H_2SO_4(aq) \rightarrow$
   $Al_2(SO_4)_3(aq) + 3H_2(g)$

   1 mark for correct formulae, **1 mark** for balancing, **1 mark** for state symbols

(c) At the start, the acid reacts with the aluminium oxide layer (1). Once this has reacted / been removed, the acid can react with the aluminium itself. (1)

133. Metal displacement reactions

1 (a) Copper is more reactive than silver. (1)

(b) $Cu(s) + 2AgNO_3(aq) \rightarrow$
   $2Ag(s) + Cu(NO_3)_2(aq)$

   1 mark for correct formulae, **1 mark** for balancing, **1 mark** for state symbols

2 (a) copper (1)

(b) A metal cannot displace itself. (1)

(c) magnesium > metal X > zinc > copper

   1 mark for correct positions of magnesium and copper, **2 marks** if all correct

3 Aluminium is more reactive than iron (1) because aluminium can displace iron from its compounds / from iron oxide. (1)

134. Explaining metal reactivity

1 A cation is a **positively** charged ion. (1)

2 (a) Ca^{2+} (1)

(b) Two (1) electrons are lost from the outer shell. (1)

(c) (i) potassium (1)

   (ii) gold (1)

(d) copper / silver / gold (1)

3 (a) Zinc is more reactive than copper (1) because it loses electrons more easily. (1)

(b) (i) $Mg(s) \rightarrow Mg^{2+}(aq) + 2e^-$

   1 mark for formulae, **1 mark** for balancing

   (ii) $2H^+(aq) + 2e^- \rightarrow H_2(g)$

   1 mark for formulae, **1 mark** for balancing

135. Metal ores

1 (a) B (1)

(b) Hydrogen is flammable / could explode. (1)

2 a rock or mineral that contains a metal / metal compound (1) in amounts high enough to make extraction worthwhile (1)

3 They are very unreactive (1) so they do not react with other elements / oxygen. (1)

4 (a) $SnO_2 + 2C \rightarrow Sn + 2CO$

   1 mark for correct formulae, **1 mark** for balancing

(b) Tin oxide is reduced (1) because it loses oxygen. (1)

5 (a) $4Na + O_2 \rightarrow 2Na_2O$

   1 mark for correct formulae, **1 mark** for balancing

(b) Copper is unreactive / low down on the reactivity series. (1)

6 (a) Ti^{4+} (1)

(b) Titanium ions are reduced (1) because they gain electrons. (1)

136. Iron and aluminium

1 (a) sodium / calcium / magnesium (1)

(b) zinc / copper (1)

2 Heat with carbon. (1)

$Fe_2O_3 + 3C \rightarrow 2Fe + 3CO$ **or** $2Fe_2O_3 + 3C \rightarrow 4Fe + 3CO_2$

1 mark for correct formulae, **1 mark** for balancing

3 (a) The ions in the electrolyte must be free to move (1) so it can conduct electricity. (1)

(b) reduces the temperature / amount of energy needed (1)

(c) at the anode: $2O^{2-} \rightarrow O_2 + 4e^-$

   1 mark for correct formulae, **1 mark** for balancing

   at the cathode: $Al^{3+} + 3e^- \rightarrow Al$

   1 mark for correct formulae, **1 mark** for balancing

4 (Production of aluminium by) electrolysis uses a lot of electricity (1), which is very expensive / more expensive than using carbon (to extract iron). (1)

137. Biological metal extraction

1 (a) correct order, starting at the top of table: 3, 5, 2, 1, 4. (All 5 correct = **3 marks**, 3 correct = **2 marks**, 1 correct = **1 mark**.)

(b) Energy could be used to heat buildings / homes / to produce electricity (1), which makes money for the company (1), OR Energy could be used in the processing of copper (1), which means that the company will need less fuel / reduce its energy costs. (1)

(c) slow process / a lot of land is needed / land could be used for food crops instead (1)

2 (a) The acid may damage rivers / streams / land / rocks / living things. (1)

(b) (i) $Fe(s) + CuSO_4(aq) \rightarrow$
   $FeSO_4(aq) + Cu(s)$

   1 mark for correct formulae, **1 mark** for balancing, **1 mark** for state symbols

   (ii) Iron is more reactive than copper / forms cations more readily than copper does. (1)

   (iii) Copper is more expensive than iron, so a more valuable product is made. (1)

138. Recycling metals

1 (a) B (1)

(b) Two of the following, for **1 mark** each: dust produced; noisy; land used; wildlife loses habitat; extra traffic; landscape destroyed

2 (a) Most lead for recycling is found in batteries (**1**) so lead does not need to be sorted from scrap metal waste. (**1**)

(b) Two of the following, for **1 mark** each: conserves metal ores / limited resources; less energy needed; fewer quarries needed / saves land / landscape; less noise / dust produced

3 (a) Steel and aluminium are much more abundant than tin in the Earth's crust (**1**); tin is much more valuable than steel or aluminium. (**1**)

(b) If little metal is used each year it may not be worthwhile recycling the metal / the mass used tells you the total amount of money / energy saved. (**1**)

139. Life-cycle assessments

1 2, 1, 4, 3 (**1**)

2 (a) 240 g = 240 / 1000 = 0.240 kg (**1**); energy used = 0.240 × 16.5 = 3.96 MJ (**1**)

(b) difference in mass of a bottle = (240 − 190) / 1000 = 0.050 kg (**1**); difference in CO_2 emissions = 0.050 × 1.2 = 0.06 kg (**1**)

3 (a) PVC: producing the material; wooden: transport and installation (**1**)

(b) The PVC frame because it uses less energy (**1**); 20% of the energy / five times less energy. (**1**)

(c) Answer could include ideas such as PVC does not biodegrade / may give off harmful substances; wood is biodegradable / rots / releases carbon dioxide when it rots. (**1**)

140. Extended response – Reactivity of metals

*Answer could include the following points: (**6**)

Basic method:

- start with powdered copper, iron, zinc, copper oxide, iron oxide and zinc oxide
- mix a spatula of a metal powder with a spatula of a metal oxide powder
- put the mixture in a steel lid
- heat strongly from below
- record observations
- repeat with a different combination of metal and metal oxide

Expected results (in writing and / or as a table, as here):

	Copper oxide	Iron oxide	Zinc oxide
Copper	not done	no visible change	no visible change
Iron	reaction seen / brown coating	not done	no visible change
Zinc	reaction seen / brown coating	reaction seen / black coating	not done

Using the results:

- count the number of reactions seen for each metal
- zinc has two reactions; iron has one reaction; copper has no reactions
- order of reactivity (most reactive first): zinc, iron, copper

Controlling risks:

- use tongs because substances / apparatus is hot
- wear eye protection to avoid contact with (hot) powders
- stand back / use a safety screen / fume cupboard to avoid breathing in escaping substances / to avoid skin contact with hot powders

141. The Haber process

1 (a) A (**1**)

(b) The reaction is reversible. (**1**)

2 (a) temperature 450 °C (**1**); pressure 200 atmospheres (**1**)

(b) Iron is a catalyst (**1**); it makes the reaction happen faster. (**1**)

3 (a) none / no visible change (**1**)

(b) (i) The rates of the forward and backward reactions are the same / equal (**1**) and they continue to happen. (**1**)

(ii) They do not change / they remain constant. (**1**)

142. More about equilibria

1 (a) The position of equilibrium moves to the right (**1**) in the direction of the fewest molecules of gas / because there are 3 mol of gas on the left of the equation but only 2 mol on the right. (**1**)

(b) The position of equilibrium moves to the left (**1**) in the direction of the endothermic reaction / away from the exothermic reaction. (**1**)

(c) The position of equilibrium moves to the right. (**1**)

(d) no change (**1**)

2 (a) 200 °C **and** 1000 atm (**1**)

(b) (i) 30% (**1**)

(ii) Lower temperatures give a greater yield (**1**) but the rate of reaction is too low. (**1**)

Higher pressures give a greater yield / rate of reaction (**1**) but very high pressures are expensive / need stronger equipment / need more energy to maintain. (**1**)

143. The alkali metals

1 C (**1**)

2 Their atoms all have one electron in their outer shell. (**1**)

3 (a) $2Na(s) + 2H_2O(l) \rightarrow 2NaOH(aq) + H_2(g)$

allow multiples or fractions, e.g.

$Na(s) + H_2O(l) \rightarrow NaOH(aq) + \frac{1}{2}H_2(g)$

1 mark for correct formulae, **1 mark** for balancing, **1 mark** for state symbols

(b) The reaction is exothermic / energy is transferred to the surroundings by heating (**1**). Sodium has a low melting point. (**1**)

(c) Lighted splint (**1**) ignites the gas with a pop. (**1**)

4 They are very reactive / react with water / react with air (**1**); oil keeps water away / air away. (**1**)

5 Three of the following, for **1 mark** each: fizzing / bubbling; metal melts / forms a ball; metal floats; metal moves around; sparks produced; lilac flame; metal disappears / dissolves / explodes at the end.

6 Going down the group, the size of the atoms increases (**1**); outer electron gets further from the nucleus / becomes more shielded / Li 2.1, Na 2.8.1, K 2.8.8.1 (**1**); attraction between outer electron and nucleus decreases / outer electron lost more easily (**1**)

144. The halogens

1 B (**1**)

2 Their atoms all have seven electrons in their outer shell. (**1**)

3 chlorine: yellow-green (**1**) gas (**1**); bromine: red-brown (**1**) liquid (**1**); iodine: dark grey (**1**) solid (forms a purple vapour) (**1**)

4 Density increases down the group (**1**); answer in the range 6000–7000 kg / m^3. (**1**)

5 (a) covalent (**1**)

(b) There are intermolecular forces / forces between molecules (**1**); these are weak / need little energy to overcome. (**1**)

(c) Melting points increase (**1**) because the intermolecular forces become stronger. (**1**)

145. Reactions of halogens

1 (a) $H_2 + F_2 \rightarrow 2HF$

1 mark for correct formulae, **1 mark** for balancing

(b) C (**1**)

(c) Bromine is less reactive than chlorine (and fluorine) (**1**) so more energy must be transferred to start the reaction / bromine gains an electron less readily. (**1**)

2 (a) $2Na(s) + Cl_2(g) \rightarrow 2NaCl(s)$

1 mark for correct formulae, **1 mark** for balancing, **1 mark** for state symbols

(b) $2Fe + 3Br_2 \rightarrow 2FeBr_3$

1 mark for correct formulae, **1 mark** for balancing

(c) (i) iron(II) Fe^{2+} (**1**); iodide I^- (**1**)

(ii) $Fe + I_2 \rightarrow FeI_2$

1 mark for correct formulae, **1 mark** for balancing

3 Fluorine atoms are smaller than chlorine atoms (**1**); outer shell is closer to the nucleus / less shielded / F 2.7, Cl 2.8.7 (**1**); greater attraction between outer electrons / shell and nucleus / outer electron gained more easily. (**1**)

146. Halogen displacement reactions

1 D (**1**)

2 (a) The order of reactivity, starting with the most reactive, is chlorine, bromine, iodine (1) because chlorine displaces bromine from bromide and iodine from iodide (1) but bromine displaces only iodine from iodide. Iodine cannot displace chlorine or bromine. (1)

(b) A halogen cannot displace itself (so no reaction will be seen). (1)

(c) Iodine will displace astatine because it is above astatine (1) so iodine is more reactive than astatine. (1)

3 (a) $F_2(g) + 2I^-(aq) \rightarrow 2F^-(aq) + I_2(aq)$;

1 mark for correct formulae, **1 mark** for balancing

(b) (i) Iodide / I^- ions are oxidised (1) because they lose electrons. (1)
(ii) Fluorine / F_2 is reduced (1) because it gains electrons. (1)

147. The noble gases

1 A (1)

2 Helium has a low density so the balloon will rise in air (1); helium is inert so it will not catch fire. (1)

3 Their outer shells are full / complete. (1)

4 (a) helium (1)
(b) Melting point increases down the group (1); answer in the range $-40\,°C$ to $-20\,°C$. (1)

5 (a) (i) 2.8 (1)
(ii) 2.8.8 (1)
(b) The outer shells of their atoms are full / complete (1) so they have no tendency to gain / lose / share electrons. (1)

148. Extended response – Groups

*Answer could include the following points: (6)

Reaction:

- caesium atoms transfer electrons
- from their outer shell
- to the outer shell of fluorine atoms
- each caesium atom loses one electron
- to form a Cs^+ ion
- each fluorine atom gains one electron
- to form an F^- ion
- oppositely charged ions / Cs^+ ions and F^- ions attract each other
- ionic bonds form

Reduction and oxidation:

- caesium is oxidised
- caesium atoms lose electrons
- $Cs \rightarrow Cs^+ + e^-$
- fluorine is reduced
- fluorine molecules gain electrons
- $F_2 + 2e^- \rightarrow 2F^-$

Vigour of reaction:

- reactivity increases down group 1 / Cs loses its electrons more easily (than Li, Na, K, Rb)
- reactivity decreases down group 7 / F gains electrons more easily (than any other group 7 element)
- caesium and fluorine are most reactive / very reactive

149. Rates of reaction

1 B (1)

2 The particles must collide (1); the collision must have enough energy / the activation energy. (1)

3 (a) a substance that speeds up a reaction without altering the products (1) and is unchanged chemically (1) and in mass (1) (at the end of the reaction)
(b) provides an alternative pathway (1) with a lower activation energy (1)
(c) (i) enzyme (1)
(ii) making alcoholic drinks / wine / beer (1)

4 The powder has a larger **surface area to volume ratio** (1) so there are more frequent collisions between reactant particles (1); **not 'there are more collisions'**.

5 There are more frequent collisions between reactant particles (1); the particles / collisions have more energy (1), so a greater proportion of collisions are successful / have the activation energy or more. (1)

150. Investigating rates

1 (a) Sodium chloride solution and water are both clear and colourless (1) so you could not tell that they are being produced. (1)
(b) (i) gas syringe / upturned burette of water / upturned measuring cylinder of water (1)
(ii) The sulfur dioxide will dissolve in the water (1) so most will not escape / the volume measurement will be incorrect / inaccurate. (1)

2 (a) Water is added to make all the total volumes $50\,cm^3$. (1)
(b) 8 (1), 24 (1), 40 (1)
(c) As the concentration increases the rate increases / rate is proportional to concentration. (1)
but
Rate is directly proportional to the concentration. (2)

151. Exam skills – Rates of reaction

1 (a) all points plotted correctly $\pm\,\frac{1}{2}$ square (2); **1 mark if one error**
single line of best fit passing through all the points (1)
Do not use a ruler to join the points (apart from the last two) because a curve is required here.
(b) The mass does not change any more / the line becomes horizontal. (1) **not 'straight'**
(c) line drawn to the left of the original line (1) becoming horizontal at 0.96 g (1)

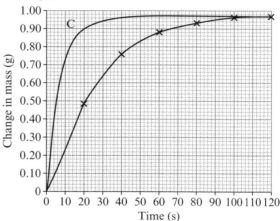

152. Heat energy changes

1 C (1)

2 In an exothermic change or reaction, heat energy is given out (1) but in an endothermic change or reaction, heat energy is taken in. (1)

3 (a) endothermic (1)
(b) acid-alkali neutralisation (1); aqueous displacement (1)

4 (a) $Mg(s) + 2HCl(aq) \rightarrow MgCl_2(aq) + H_2(g)$

1 mark for correct formulae, **1 mark** for balancing, **1 mark** for state symbols

(b) Measure the temperature of the acid before and after adding magnesium (1) using a thermometer (1) and the temperature should increase. (1)

(c) More heat energy is released (1) forming bonds in the products (1) than is needed to break bonds in the reactants. (1)

153. Reaction profiles

1 Reactants have more stored energy than products / products have less stored energy than reactants (1); difference in energies shows that heat energy is given out. (1)

2 (a) diagram completed with upwards arch between reactants and product lines (1); activation energy correctly identified (1)

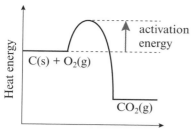

(b) the minimum amount of energy needed to start a reaction (1)

3 diagram complete with reactant line below products line (1); upward arch between reactant and products lines (1); activation energy correctly identified (1); overall energy change correctly identified (1)

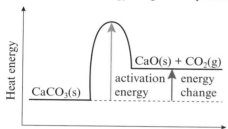

154. Calculating energy changes

1 (a) (1 × 436) + (1 × 243) = 436 + 243
= 679 kJ mol^{-1} (1)
(b) (2 × 432) = 864 kJ mol^{-1} (1)
(c) energy change = (energy in) − (energy out) = 679 − 864 kJ mol^{-1} (1)
= −185 kJ mol^{-1} (1)
(d) exothermic (1) because the energy change is negative / more energy is given out than is taken in (1)

2 energy in = (1 × 945) + (3 × 436)
= 945 + 1308
= 2253 kJ mol^{-1} (1)

energy out = (6 × 391) = 2346 kJ mol^{-1} (1)

energy change = 2253 − 2346 kJ mol^{-1} (1)
= −93 kJ mol^{-1} (1)

155. Crude oil

1 D (1)

2 A finite resource is no longer being made / is made extremely slowly (1); crude oil takes millions of years to form. (1)

3 (a) C_6H_{14} (1)
(b) They are compounds of hydrogen (1) and carbon only. (1)

4 (a) $C_8H_{18} + 12\frac{1}{2}O_2 \rightarrow 8CO_2 + 9H_2O$ or
$2C_8H_{18} + 25O_2 \rightarrow 16CO_2 + 18H_2O$

1 mark for correct formulae, **1 mark** for balancing

(b) Bubble gas through limewater (1), which turns milky / cloudy white. (1)

5 a starting material (1) for an industrial chemical process (1)

156. Fractional distillation

1 (a) D (1)
(b) (i) bitumen (1)
(ii) kerosene (1)
(c) petrol (1); diesel oil (1)

2 alkanes (1)

3 The viscosity of fuel oil is too high for the fuel to flow easily (1) and it does not vaporise easily because its boiling point is too high. (1)

4 There are weak (1) forces between molecules / intermolecular forces (1) so only a little energy needed to overcome / break these forces. (1)
not covalent bonds

5 Oil is evaporated (1) and passed into a column, which is hot at the bottom and cool at the top (1); hydrocarbons (rise) cool and condense

at different heights (1), depending on boiling point / size of molecules / strength of intermolecular forces. (1)

157. Alkanes

1 C (1)

2 (a) C_nH_{2n+2} (1)
(b) (i) $C_{12}H_{26}$ (1)
(ii) carbon dioxide (1); water (1)
(iii) The alkanes react with oxygen / oxides are formed / carbon atoms and hydrogen atoms gain oxygen. (1)

3 (a) Two from the following, for **1 mark** each: all contain carbon / hydrogen / oxygen / O–H group / covalent bonds / are simple molecules
(b) differ by CH_2 / one carbon atom **and** two hydrogen atoms (1)

158. Incomplete combustion

1 A (1)

2 (a) When breathed in, carbon monoxide combines with haemoglobin / red blood cells (1) so less oxygen can be carried / there is a lack of oxygen to cells. (1)
(b) They cause lung disease / bronchitis / make existing lung disease worse. (1)

3 Three from the following, for **1 mark** each: nest restricts entry of oxygen; carbon monoxide will be produced; soot will be produced; carbon monoxide is toxic; soot causes breathing problems

4 (a) $C_3H_8 + 5O_2 \rightarrow 3CO_2 + 4H_2O$
1 mark for correct formulae, **1 mark** for balancing
(b) $C_3H_8 + 3\frac{1}{2}O_2 \rightarrow 4H_2O + C + CO + CO_2$
1 mark for correct formulae, **1 mark** for balancing

159. Acid rain

1 (a) $N_2 + 2O_2 \rightarrow 2NO_2$
1 mark for correct formulae, **1 mark** for balancing
(b) Oxygen and nitrogen from the air (1) react together at the high temperatures inside the engine. (1)
(c) (i) sulfur (1)
(ii) sulfur reacts with oxygen in the air (1)
$S + O_2 \rightarrow SO_2$ (1)

2 $SO_2(g) + H_2O(l) \rightarrow H_2SO_3(aq)$
1 mark for correct formulae and balancing, **1 mark** for state symbols

3 (a) Marble / calcium carbonate reacts with acids / acidic rainwater (1) but granite does not. (1)
(b) damage to trees / plants / soil (1); makes lakes acidic / harms aquatic life (1)

160. Choosing fuels

1 It is being used up faster than it can form. (1)

2 (a) crude oil (1)
(b) Petrol is used as a fuel for cars (1). Kerosene is used as a fuel for aircraft (1). Diesel oil is used as a fuel for some cars / some trains. (1)

3 (a) $2H_2 + O_2 \rightarrow 2H_2O$
1 mark for correct formulae, **1 mark** for balancing
(b) carbon dioxide (1)

4 (a) volume needed = 12 000 × (100 / 141.8)
= 8460 dm^3 (1)
(b) volume needed = 1.36 × (100 / 47.3)
= 2.88 dm^3 (1)
(c) (i) Hydrogen releases more energy per kg than petrol. (1)
(ii) A much smaller volume of petrol is needed to release the same amount of energy as hydrogen / petrol is more energy dense. (1)

161. Cracking

1 D (1)

2 (a) alkene / unsaturated (1)
(b) $C_{10}H_{22} \rightarrow C_8H_{18} + C_2H_4$
1 mark for correct reactant, **1 mark** for correct products

3 a reaction in which larger alkanes are broken down into smaller (more useful) alkanes / smaller (more useful) saturated hydrocarbons **(1)** and smaller alkenes / unsaturated hydrocarbons **(1)**

4 (a) to make smaller molecules **(1)** that are in higher demand / more useful **(1)**
  (b) (from well B) because most of the alkanes have a small number of carbon atoms in their molecules **(1)** so are in demand / more useful already **(1)**

162. Extended response – Fuels

*Answer could include the following points: **(6)**

Why incomplete combustion happens:

- incomplete combustion happens when there is insufficient oxygen / air
- this can happen if there is not enough ventilation, such as inside a tent
- not enough oxygen for complete combustion

Products and their problems:

- carbon monoxide gas produced
- carbon monoxide is toxic
- combines with haemoglobin / red blood cells
- so less oxygen can be carried / there is a lack of oxygen to cells
- can cause unconsciousness / death
- carbon particles / soot produced
- cause lung disease / bronchitis / make existing lung disease worse
- cause blackening, e.g. of the bottom of the kettle / the inside of the tent
- balanced equation, e.g. $C_3H_8 + 3O_2 \rightarrow 4H_2O + 2CO + C$; **many are possible**

Other problems:

- less energy is released by incomplete combustion
- wastes camping gas / fuel
- takes longer to heat the water / to make the tea / to cook

163. The early atmosphere

1 B **(1)**

2 (a) nitrogen **(1)**
  (b) Water vapour (in the atmosphere) **(1)** condensed (and fell as rain) **(1)**.
  (c) Carbon dioxide dissolved **(1)** in the oceans / water **(1)**.

3 A **glowing** splint **(1)** relights. **(1)**

4 The growth of primitive plants used carbon dioxide **(1)** and released oxygen **(1)** via photosynthesis **(1)**.

5 They use different evidence / draw different conclusions from the same evidence / study different parts of the Earth / no direct measurements are possible / no humans were on Earth at the time. **(1)**

164. Greenhouse effect

1 (a) methane **(1)**
  (b) (i) Fossil fuels contain hydrocarbons / carbon **(1)**, which react with oxygen in the air to produce carbon dioxide. **(1)**
     (ii) increasing use of fossil fuels **(1)** so carbon dioxide is released faster **(1)** than it can be removed (e.g. by photosynthesis, dissolving in the oceans) **(1)**

2 Various gases in the atmosphere, such as carbon dioxide, absorb heat radiated from the Earth **(1)** and then release energy / heat energy (in all directions) **(1)** which keeps the Earth warm. **(1)**

3 (a) The measurements could not be taken directly / they are historical / their location may not be representative of the whole planet. **(1)**
  (b) As the carbon dioxide level increases, the global temperature increases **(1)**; as the carbon dioxide level decreases, the global temperature decreases. **(1)**
  (c) climate change / change in global weather patterns / ice caps melting / sea level rises / loss of habitats **(1)**

165. Extended response – Atmospheric science

*Answer could include the following points: **(6)**

Greenhouse effect:

- carbon dioxide and some other gases in the atmosphere absorb heat energy
- radiated from the Earth
- then release energy
- which keeps the Earth warm

Processes releasing carbon dioxide:

- burning fossil fuels
- respiration
- volcanic activity

Processes absorbing carbon dioxide:

- dissolving in seawater
- photosynthesis

Discussing the data:

- as the concentration of carbon dioxide rises
- the mean global temperature rises
- human activity, e.g. burning fossil fuels, could cause increase in temperature
- but there are some years when the temperature decreases
- carbon dioxide is also produced by other processes
- so it might not all be due to human activity
- there might be a common factor not shown in the graphs that is responsible for both changes

Physics

166. Key concepts

1 All eight for **2 marks**, any four for **1 mark**: ampere, A; joule, J; pascal, Pa; coulomb, C; mole, mol; watt, W; newton, N; ohm, Ω.

2 A base unit is independent of any other unit **(1)**; a derived unit is made up from two or more base units. **(1)**

3 (a) 0.75 kg **(1)** (b) 750 W **(1)** (c) 1500 s **(1)**
  (d) 0.03 m **(1)** (e) 3 000 000 J **(1)**

4 (a) 2.5 kHz = 2500 Hz **(1)**; 2.5×10^3 Hz **(1)**
  (b) 8 nm = 0.000000008 m **(1)**; 8×10^{-9} m **(1)**

5 $s = d \div t = 75$ m $\div 10.5$ s $= 7.142\,857\,142\,857\,143$ m / s **(1)** so to 5 significant figures = 7.1430 m / s **(1)**

167. Scalars and vectors

1 (a) scalars: speed, energy, temperature, mass, distance **(1)**; vectors: acceleration, displacement, force, velocity, momentum **(1)**
  (b) (i) any valid choice and explanation, e.g. mass **(1)** is a scalar because it has a size / magnitude but no direction **(1)**
     (ii) any valid choice and explanation, e.g. force **(1)** is a vector because it is has a magnitude and direction **(1)**

2 (a) Velocity is used because both a size and a direction are given **(1)** and the swimmers are swimming in different directions. **(1)**
  (b) The second swimmer is swimming in the opposite direction to the first swimmer. **(1)**

3 (a) D **(1)**
  (b) Weight has a size / magnitude and a direction but all the other quantities just have a magnitude. **(1)**

4 Both the wind speed and the speed of the aeroplane have direction as well as magnitude and so are vectors **(1)**. As they have different directions **(1)**, the pilot needs to take this into account when planning the route and this will also affect the time taken to fly the route. **(1)**

168. Speed, distance and time

1 (a) (i) B **(1)**
     (ii) C **(1)**
  (b) In part A he travels 60 m in 40 s. Speed = distance ÷ time **(1)** = 60 m ÷ 40 s **(1)** = 1.5 m / s **(1)** (You can use any part of the graph to read off the distance and the time as the line is straight; you should always get the same speed).
  (c) Displacement is the length and direction of a straight line between the runner's home and the park **(1)** but the distance the runner ran probably included bends and corners on the path the runner took. **(1)**

2 (a) speed = 84 m ÷ 24 s **(1)** = 3.5 **(1)** m / s **(1)**
  (b) 3.5 m / s upwards or up **(1)**

3 time = distance ÷ speed = 400 m ÷ 5 s **(1)** = 80 s **(1)**

169. Equations of motion

1 All four correct for **2 marks**; two or three correct for **1 mark**. *a*: acceleration; *x*: distance; *v*: final velocity; *u*: initial velocity

2 (a) $a = (v − u) ÷ t$ **(1)** = (25 m / s − 15 m / s) ÷ 8 s **(1)** = 1.25 m / s² **(1)**
(b) $v^2 = u^2 + 2 × a × x$ = (25 m / s)² + 2(1.25 m / s² × 300 m) **(1)** = 1375 m² / s² **(1)** $v = \sqrt{1375}$ m / s = 37 m / s **(1)** (allow rounding error: answers between 37.00 m / s and 37.10 m / s)
(c) $v^2 − u^2 = 2 × a × x$ so $x = (v^2 − u^2) ÷ (2 × a)$ **(1)** = ((5 m / s)² − 1375 m² / s²) ÷ (2 × −2 m / s²) **(1)** = −1350 m² / s² ÷ −4 m / s² = 337.5 m **(1)**

170. Velocity / time graphs

1 (a) $a = (v − u) ÷ t$ = (4 − 0) ÷ 5 **(1)** = 0.8 **(1)** m / s² **(1)**
(b) Plot a velocity / time graph **(1)**; the total area under the line can be calculated **(1)** and this gives the total distance travelled. **(1)**

2 (a) A **(1)** C **(1)**
(b) a right-angled triangle with a horizontal and vertical side that covers as much of the line as possible **(1)**
(c) change in velocity = 30 m / s, time taken for change = 5 s; acceleration = (change in velocity) ÷ (time taken) = 30 m / s ÷ 5 s **(1)** = 6 m / s² **(1)** (the triangle drawn may be different but the answer should be the same)
(d) area under line = ½ × 5 s × 30 m / s **(1)** = 75 m **(1)**

171. Determining speed

1 Commuter train: 55 m / s; speed of sound in air: 330 m / s; walking: 1.5 m / s. **(1)** All three needed for **1 mark**.

2 (a) The light beam is cut / broken by the card as it enters the light gate and this starts the timer **(1)**. When the card has passed through, and the light beam is restored **(1)** this stops the timer. **(1)**
(b) Speed is calculated from the length of the card and the time taken for the card to pass through the light gate. **(1)**

3 Very short distances may be measured using this method and this gives a good measure of instantaneous speed **(1)**. Short times are difficult to measure accurately with a stopwatch **(1)**. Errors associated with human error / reaction time / parallax are reduced. **(1)**

172. Newton's first law

1 Four arrows drawn: vertical: down = weight, up = upthrust (arrows the same length); horizontal left to right = thrust or force from the engines, right to left = water resistance or drag, the thrust arrow should be longer than the drag arrow (**1 mark** for all the forces correctly named and **1 mark** for the corresponding relative lengths of the arrows).

2 (a) The action is the downwards force of the skater on the ice **(1)** and the reaction is the (upwards) force of the ice on the skater. **(1)** (Can be the other way around).
(b) resultant force = 30 N − 10 N − 1 N **(1)** = 19 N **(1)**
(c) The resultant force is zero / 0 N **(1)** so the velocity is constant / stays the same. **(1)**
(d) Only the resistance forces are acting now, so the resultant force is backwards / against the motion of the skater **(1)**, so the skater slows down / has negative acceleration. **(1)**

3 (a) Assume downwards is positive; so resultant downward force = 1700 N − 1900 N = − **(1)** 200 N **(1)** (State which direction you are using as the positive direction.)
(b) The velocity of the probe towards the Moon will decrease **(1)** because the force produces an upwards acceleration / negative acceleration. **(1)**

173. Newton's second law

1 (a) The trolley will accelerate **(1)** in the direction of the pull / force. **(1)**
(b) The acceleration is smaller / lower **(1)** because the mass is larger. **(1)**

2 (a) $F = m × a$ = 3000 kg × −13 m / s² **(1)** = −39 000 **(1)** N **(1)**
(b) in the opposite direction to the spacecraft's motion / upwards **(1)**

3 (a) $a = F ÷ m$ = 10 500 N ÷ 640 kg **(1)** = 16.4 **(1)** m / s² **(1)**
(b) The mass of the car decreases **(1)** so the acceleration will increase. **(1)**

174. Weight and mass

1 The mass of the LRV on the Moon is 210 kg **(1)** because the mass of an object does not change if nothing is added or removed. **(1)**

2 B **(1)**

3 $W = m × g$ **(1)** so (1 + 2 + 1.5) kg × 10 N / kg = 4.5 kg × 10 N / kg **(1)** = 45 N **(1)**

4 calculating correct masses for all three items **(1)**; selecting correct items **(1)**; clothes 10.5 kg + camera bag 5.5 kg + jacket 3.5 kg = 19.5 kg **(1)**

175. Force and acceleration

1 Electronic equipment is much more accurate **(1)** than trying to obtain accurate values for distance and time to calculate velocity, then calculate acceleration **(1)**, when using a ruler and a stopwatch. (Reference should be made to distance, time and velocity.)

2 Acceleration is inversely proportional to mass. **(1)**

3 Acceleration is calculated by the change in speed ÷ time taken, so two velocity values are needed **(1)**; the time difference between these readings **(1)** is used to obtain a value for the acceleration of the trolley.

4 (a) For a constant slope, as the mass increases, the acceleration will decrease **(1)** due to greater inertial mass. **(1)**
(b) Newton's second law, $a = F ÷ m$ **(1)**

5 An accelerating mass of greater than a few hundred grams can be dangerous and may hurt somebody if it hits them at speed **(1)**. Any two of the following precautions: do not use masses greater than a few hundred grams **(1)**, wear eye protection **(1)**, use electrically tested electronic equipment **(1)**, avoid trailing electrical leads. **(1)**

176. Circular motion

1 (a) The velocity of an orbiting satellite changes because the direction is constantly changing **(1)** even though speed remains constant. **(1)**
(b) The force provided by the Earth's gravitational field causes it to change direction as it orbits the Earth at constant speed. **(1)**

2 B **(1)**

3 tension **(1)**, any valid example of a rope / string and mass, etc. **(1)**; gravitational **(1)**, any valid example of an orbiting mass **(1)**; frictional **(1)**, any circling body with dynamic friction **(1)**

4 (a) centripetal force **(1)**
(b) The ball is constantly changing direction **(1)** (due to the centripetal force acting on it).

177. Momentum and force

1 B **(1)**

2 (a) Force is the rate of change of momentum **(1)**. It is the change in momentum divided by the time taken for the change. **(1)**
(b) $p = m × v$ **(1)** = 1500 kg × 25 m/s **(1)** = 37 500 kg m/s, so change in momentum = 37 500 kg m/s **(1)**
(c) The forces exerted on the passenger are large when the mass is large **(1)** or the deceleration of the vehicle is large **(1)**. Fitting an airbag / crumple zone / seat belt **(1)** increases the time over which a passenger comes to rest. So this will reduce the force exerted on them. **(1)**

3 $F = (mv − mu) ÷ t$ = ((500 kg × 15 m/s) − (500 kg × 10 m/s)) **(1)** ÷ 20 s = (7500 − 5000) kg m/s ÷ 20 s **(1)** = 125 N **(1)**

4 Three from: The hockey player should try to make sure that the change in speed between the hockey stick and the ball is as high as possible / move the hockey stick very quickly **(1)**. The contact time between the club and ball needs to be as small as possible / the ball needs to be hit quickly **(1)**. The large change in velocity **(1)** and the small contact time **(1)** make the force to move the ball as large as possible. **(1)**

178. Newton's third law

1 D **(1)**

2 momentum = mass × velocity **(1)** = 1200 kg × 30 m/s **(1)**; momentum = 36 000 kg m/s **(1)** in the south direction **(1)**

3 (a) momentum of Dima and car = 900 kg × 1.5 m/s = 1350 **(1)** kg m/s
(b) (i) momentum of Sam and car = 900 kg × 3 m/s = 2700 **(1)** kg m/s
(ii) The total momentum of the two cars after the collision must equal the total momentum of the two cars before the collision **(1)** so total momentum of both cars is unchanged. **(1)**
(iii) total momentum = 1350 + 2700 = 4050 kg m/s **(1)** so velocity after collision = total momentum ÷ total mass = 4050 kg m/s ÷ 1800 kg **(1)** = 2.25 m/s **(1)**

Answers

4 momentum of skater 1 before collision = 50 kg × 7.2 m/s = 360 kg m/s; momentum of skater 2 before collision = 70 kg × 0 m/s = 0 kg m/s; momentum of both skaters after collision = 360 + 0 = 360 kg m/s **(1)** so combined velocity = 360 kg m/s ÷ (70 + 50) kg **(1)** = 3 m/s **(1)**

179. Human reaction time

1 B **(1)**

2 Human reaction time is the time between a stimulus occurring and a response **(1)**. It is related to how quickly the human brain can process information and react to it. **(1)**

3 A person sits with their index finger and thumb opened to a gap of about 8 cm **(1)**. A metre ruler is held, by a partner, so that it is vertical and exactly level with the person's finger and thumb, with the lowest numbers on the ruler at the bottom **(1)**. The ruler is dropped and then grasped by the other person. **(1)**

4 (a) 0.20–0.25 s **(1)**

(b) Success in certain professions relies on short reaction times, where a fast response could result in ensuring the safety of others **(1)** or is necessary for a competitive career, such as sports **(1)**. Any two suitable examples with justification, e.g. an international tennis player who is able to react to the opponent hitting the ball **(1)** or a racing driver who can react quickly to changing situations. **(1)**

5 Time = distance ÷ speed so time = 25 ÷ 20 **(1)** Reaction time = 1.25 **(1)** (s).

180. Stopping distances

1 (a) Thinking distance **(1)**; braking distance. **(1)**

(b) 1. Thinking distance; 2. Braking distance **(1)** (both needed for mark)

(c) *Thinking distance will increase if*: the car's speed increases, the driver is distracted, the driver is tired, the driver has taken alcohol or drugs **(1)** (all four needed for mark). *Braking distance will increase if*: the car's speed increases, the road is icy or wet, the brakes or tyres are worn, the mass of the car is bigger **(1)** (all four needed for mark).

(d) Worn tyres will have less surface area in contact with the road than new tyres, so frictional force opposing motion will be less **(1)**; therefore the car will take longer to stop than when the tyres were new. **(1)** (This affects the braking distance component. The thinking distance component remains unchanged.)

2 Using a mobile phone affects thinking distance **(1)**. This is because it causes a distraction to the driver / affects reaction time. **(1)**

3 Driving faster will increase thinking distance **(1)** and braking distance **(1)**. If drivers do not increase their normal distance behind the vehicle in front accordingly there is an increased risk of an accident / collision. **(1)**

181. Extended response – Motion and forces

*Answer could include the following points: **(6)***

- Acceleration is the rate of change of velocity (speed in a given direction) so although the speed is constant, the direction is continually changing for an object in circular motion.
- For motion in a circle there must be a resultant force, known as a centripetal force, which acts towards the centre of the circle.
- The string represents the centripetal force, which acts towards the centre of the circle.
- Extend the investigation with different lengths of string.
- Extend the investigation with different masses.
- Improve data collection with electronic sensors.
- Improve data analysis with video or photography.
- Reference to the importance of control variables for valid data collection.

182. Energy stores and transfers

1 B **(1)**

2 The energy transfer diagram shows that the total amount of energy in the stores before the transfer is equal to the total amount of energy in the stores after the transfer **(1)** so there is no net change, supporting the conservation of energy. **(1)**

3 chemical store **(1)**; kinetic store and thermal store **(1)**; thermal store **(1)**

4 (a) Gravitational store bar is lower **(1)**; kinetic store bar higher **(1)**; rise in kinetic store bar should equal the amount down in the gravitational store bar. **(1)**

(b) Chemical store bar is lower **(1)** than kinetic store bar **(1)**; thermal store bar higher **(1)**; total heights of all bars should be the same as the total heights of the bars before use. **(1)**

183. Efficient heat transfer

1 concrete **(1)**: this has the lowest relative thermal conductivity, which means it will have the slowest rate of transfer of thermal energy. **(1)**

2 (a) Thicker walls provide more material for the thermal energy to travel through from the inside to the outside **(1)** so the rate of thermal energy loss is less, keeping the houses warmer. **(1)**

(b) Thicker walls provide more material for the thermal energy to travel through from the outside to the inside **(1)** so the rate of thermal energy transfer is less, keeping the houses cool. **(1)**

3 useful energy transferred = energy transferred to the box = 1 000 000 J; total energy used by the crane = the energy stored in the fuel = 4 000 000 J; efficiency = 1 000 000 J ÷ 4 000 000 J × 100 **(1)** = 25% **(1)** (or calculation can omit × 100 and leave efficiency as 0.25)

4 (a) efficiency = 20% ; thermal (wasted) energy = 80% **(1)** so this is 4 × 40 = 160 J **(1)**; so total energy in = 40 + 160 = 200 J **(1)**

(b) 200 W **(1)**

184. Energy resources

1 (a) hydroelectric; geothermal **(1)**

(b) Demand is greatest at certain times of the day **(1)**. Demand may be high when some renewable sources may not be available. **(1)**

2 (a) A hydroelectric power station is a reliable producer of electricity because it uses the gravitational potential energy of water which can be stored until it is needed **(1)**. As long as there is no prolonged drought / lack of rain the supply should be constant. **(1)**

(b) Any of the following: Hydroelectric power stations have to be built in mountainous areas / high up (compared to supply areas so that the gravitational potential energy can be captured) **(1)**. The UK has very few mountainous areas like this **(1)**. It is limited to areas such as north Wales and the Scottish Highlands. **(1)**

3 (a) When carbon dioxide is released into the atmosphere it contributes to the greenhouse effect / build-up of CO_2 **(1)**, which is believed to contribute to global warming. **(1)**

(b) Sulfur dioxide and nitrogen oxides have been found to dissolve in the water droplets in rain clouds, increasing their acidity **(1)**; this can kill plants / damage forests and lakes / dissolve the surfaces of historical limestone buildings. **(1)**

(c) Fossil fuel power stations do not rely on the energy stores in the environment **(1)** and so can be built in re-developed areas / do not need to be positioned to take advantage of wind / tidal / wave / hydroelectric resources. **(1)**

4 Two from: We do not know how much oil there is to extract **(1)**, how fast it can be extracted **(1)** or how demand for oil may change. **(1)**

185. Patterns of energy use

1 (a) After 1900 the world's energy demand rose as the population grew **(1)**. There was development in industry which increased the demand for energy **(1)** and the rise of power stations using fossil fuels added to demand. **(1)**

(b) (i) non-renewable energy resources / fossil fuels / oil, coal, natural gas **(1)**

(ii) Any two from: population has increased so energy consumption is higher **(1)**; industrial / technological developments require more energy **(1)**; transport networks have grown **(1)** (any other valid reason)

(iii) Nuclear research began in the 1940s. **(1)**

(iv) hydroelectric **(1)**

2 Any six points made for **6 marks**: As the population continues to rise the demand for energy will also continue to rise **(1)**. Current trends show the use of fossil fuels being a major contributor to the world's energy resources **(1)**. These are running out and no other energy resource has so far taken their place **(1)**. This could lead to a large gap between demand and supply **(1)**. To match the rise in demand for energy, further research and development of non-renewable resources will need to be made (1) to provide reliable (1) and cost-effective (1) energy supplies. While cheaper fossils fuels still remain available there is less incentive for governments to do this. (1)

186. Potential and kinetic energy

1   D (1)

2   Kinetic energy = ½ × m × v^2 = ½ × 70 × 6^2 (1) = 1260 (1) J

3   (a) ΔGPE = 2000 kg × 10 N / kg × 0.5 m (1) = 10 000 J (1)
    (b) 10 000 J (or same answer as given to part (a)) (1)
    (c) KE = work done or GPE gained = 10 000 J (1); v^2 = KE ÷ (0.5 × m) (1)
       so v^2 = 10 000 ÷ 1 000 (1) = 10
       so v = √10 = 3.16 m / s (1)

4   *The indicative content below is not prescriptive and you are not required to include all the material which is indicated as relevant. Additional content included in the response must be scientific and relevant.*

    Four from: Kinetic energy (1) transferred to the ball reduces (1) as it climbs to the top of the curve where KE is a minimum (1) and gravitational potential energy reaches maximum (1). Some of the kinetic energy transferred to the ball is dissipated to the surroundings (1) as thermal energy (1) due to air resistance / drag / friction. (1)

187. Extended response – Conservation of energy

*Answer could include the following points: (6)

- Refer to the change in gravitational potential energy (GPE) as the swing seat is pulled back / raised higher.
- Before release, the GPE is at maximum / kinetic energy (KE) of the swing is at a minimum.
- When the swing is released, the GPE store falls and the KE store increases.
- KE is at a maximum at the mid-point, GPE is at a minimum.
- The system is not 100% efficient; some energy is dissipated to the environment.
- Friction due to air resistance and / or at the pivot results in the transfer of thermal energy to the surroundings / environment.
- Damping, due to friction, will result in the KE being transferred to the thermal energy store of the swing and hence to the environment.
- Eventually all the GPE will have been dissipated to the surroundings / environment (so is no longer useful).

188. Waves

1   Sound waves are this type of wave: L; All electromagnetic waves are this type of wave: T; Particles oscillate in the same direction as the wave: L; They have amplitude, wavelength and frequency: B; Seismic S waves are this type of wave: T; They transfer energy: B. All six correct – **3 marks**; five correct – **2 marks**; three correct – **1 mark**.

2   (a) B (1)
    (b) 6 cm / 0.06 m (1)
    (c) any correct wave with higher amplitude (1) and shorter wavelength (1)

3   (a) When a sound wave is generated each particle oscillates (1) in the same direction as the direction in which the wave travels. (1)
    (b) When a water wave is generated the surface particles oscillate (1) at 90° / perpendicular to the direction in which the wave travels. (1)

189. Wave equations

1   distance travelled by the waves (in metres) = 30 000 m (1); time taken = 20 s; speed of sound = 30 000 m ÷ 20 s (1) = 1500 m / s (1)

2   wave speed = 0.017 m × 20 000 Hz (1) = 340 m / s (1)

3   λ = v ÷ f = 0.05 m / s ÷ 2 (1) = 0.025 (1) m (1)

4   x = v × t (1) = 300 000 000 m / s × 0.12 s (1) = 36 000 km (1)

190. Measuring wave velocity

1   frequency of the waves (f) = 3 Hz; wavelength of the waves (λ) = 0.05 m; speed of waves = 3 Hz × 0.05 m (1) = 0.15 (1) m / s (1)

2   D (1)

3   *Use of* v = x ÷ t, so v = 50 ÷ 15 (1) = 3.3 (1) m / s (to 2 significant figures)

4   (a) They can process the data using the equation v = x ÷ t (1)
    (b) Use an electronic data collector (1); repeat the experiment at 50 m (1); repeat the experiment over a range of distances. (1)

191. Waves and boundaries

1   refraction (1); normal (1); do not (1)

2   Reflection: the wave bounces back at a surface but does not pass through. Refraction: the wave passes through but at a changed speed. Absorption: the wave energy is transferred into a thermal energy store. All three correct: **2 marks**; two correct: **1 mark**.

3   Sound waves are generated by vibrating surfaces that push against the air particles (1). The air particles oscillate / move back and forth (1) transferring energy through the air in the form of pressure waves. (1)

4   (a) When waves move from deeper water to shallower water they slow down (1), the waves become closer together (1) and they change direction. (1)
    (b) The change in depth of water represents the boundary / interface between the two media through which light travels. (1)
    (c) The change in direction (of both water waves and light waves) is caused by the change in speed of the waves. (1)

192. Waves in fluids

1   (a) Count the number of waves that pass a point each second and do this for one minute (1); divide the total by 60 to get a more accurate value for the frequency of the water waves. (1)
    (b) Use a stroboscope to 'freeze' the waves (1) and find their wavelength by using a ruler in the tank / on a projection. (1)
    (c) wave speed = frequency × wavelength or v = f × λ (1)
    (d) the depth of the water (1)

2   A ripple tank can be used to determine a value for the wavelength, frequency and wave speed of water waves (1), as long as small wavelengths (1) and small frequencies are used. (1)

3   water: hazard – spills may cause slippages; safety measure – report and wipe up immediately (1); electricity: hazard – may cause shock / trailing cables may form trip hazard; safety measure – do not touch plugs / wires / switches with wet hands / keep cables tidy (1); strobe lamp: hazard – flashing lights may cause dizziness or fits; safety measure – check that those present are not affected by flashing lights (1)

193. Extended response – Waves

*Answer could include the following points: (6)

- The wave behaviour shown is refraction (from the key to the man). This is a property of waves.
- Light is a wave and so is refracted through transparent or translucent materials.
- When light passes from one material to another of different density it is refracted at the boundary due to a change of speed.
- When light passes from a more dense material to a less dense material it is refracted away from the normal (as in this case).
- Actual position of the key is at position A.
- The key appears to be at position B because the man's brain extrapolates the refracted wave (as shown by the dashed line).

194. Electromagnetic spectrum

1   B (1)

2   (a) All parts of the electromagnetic spectrum are transverse waves (1) and they all travel at $3 × 10^8$ m / s / the same speed in a vacuum. (1)
    (b) The different waves carry different amounts of energy. (1)

3   A: X-rays (1); B: visible light (1); C: microwaves (1)

4   v = f × λ, so f = v ÷ λ (1) = $3 × 10^8$ m / s ÷ 240 m (1) = $1.25 × 10^6$ Hz (1)

195. Investigating refraction

1   (a) Four from: Place a refraction block on white paper and connect a ray box to an electricity supply (1); switch on the ray box and set it at an angle to the surface of the block (1); use a sharp pencil to draw around the refraction block and make dots down the centre of the rays either side of the block (1); use a sharp pencil and ruler to join the 'external' rays and then draw a line across the outline of the block to join the lines (1); use a protractor to draw a normal where the light ray met the block and measure the angle of incidence and angle of refraction. (1)

(b) When a light ray travels from air into a glass block, its direction changes (1) and the angle of refraction will be less than the angle of incidence. (1)

(c) (i) The ray of light would not change direction. (1)

(ii) The light would slow down (1) (travelling from a less dense medium to a more dense medium) and the wave fronts would be closer together / the wavelength of light would be shorter. (1)

2 Three from: use of electricity: if mains electricity is used there is a risk of shock – use tested apparatus / do not try to plug in / unplug in the dark (1); experiments are generally done in low-level light so there is a risk of tripping – clear floor area and working space (no trailing wires) and avoid moving around too much (1); if glass blocks are used there is a risk of cuts – handle with care, use Perspex / non-breakable blocks when possible (1); ray boxes get hot so risk of burns – do not touch ray boxes during operation. (1)

3 The waves travel more slowly and wavelength becomes shorter in shallow water (1); the waves change direction / bend towards the 'normal' as they move into shallower water. (1)

196. Wave behaviour

1 C (1)

2 (a) Reflection: waves bounce off a surface (1); refraction: waves change speed and direction when passing from one material to another (1); transmission: electromagnetic waves are transmitted when they pass through a material (1); absorption: different electromagnetic waves are absorbed by different materials. (1)

(b) Two valid examples, e.g. reflection: light on a mirror (1); refraction: light through water (1); transmission: radio waves passing through the atmosphere from transmitter to receiver (1); absorption: X-rays absorbed by the atmosphere. (1)

3 (a) Microwaves are shorter in wavelength (1) and higher in frequency (1) than radio waves.

(b) Microwaves sent from the ground transmitter are able to pass through the ionosphere (1) and are received and re-emitted by the receiver to the ground (1). Radio waves sent from the ground transmitter are refracted by the ionosphere (1) and then reflected back to the receiver on the ground. (1)

4 As **oscillating** charges move up and down a radio aerial oscillating **electric** and magnetic fields move from the antenna, across space (**1 mark** – both words needed). When the oscillating electric **field** encounters another aerial, it causes oscillations in the receiving **electrical** circuits (**1 mark** – both words needed).

5 Space-based telescopes are outside the Earth's atmosphere (1) so they are able to detect the whole range of electromagnetic waves (1) that are emitted by stars and galaxies. (1)

197. Dangers and uses

1 (a) A and C (1)

(b) B and D (1)

(c) A and D (1)

2 Some electromagnetic waves can be dangerous. **Microwaves** can **heat** the water inside our bodies causing significant damage to cells. **Infrared** waves transfer **thermal** energy and can cause burns to skin. **Ultraviolet** waves can damage **eyes** and can cause skin cancer. **1 mark** for each correct sentence.

3 X-rays and gamma rays can cause damage to DNA in cells / produce free radicals that can damage DNA (1). This may lead to cell death or cause cancer. (1)

4 Three from: X-rays are useful because they can be used to diagnose injuries without surgery (1) / check for medical conditions without surgery (1). X-rays can be harmful to cells in the body (1) / can damage DNA in cells (1). The use of X-rays should be controlled by carefully recording the number of X-rays delivered over time to prevent overexposure (1) / using ultrasound scans when risk is high (e.g. foetal scanning) to prevent over exposure. (1)

198. Changes and radiation

1 D (1)

2 (a) Energy depends on frequency (1) and the higher the frequency, the greater the energy carried by the wave. (1)

(b) When an electron absorbs electromagnetic radiation it moves up one or more energy levels in the atom or may leave the atom entirely. (1)

(c) When an electron emits electromagnetic radiation it moves down one or more energy levels in the atom. (1)

3 When an electron is 'excited', it has absorbed enough energy to move to one of the higher energy levels, e.g. $n = 1$ to $n = 3$ (1). The electron then returns to a lower energy level, emitting a photon of the same / equivalent energy (1) as the difference between the energy levels between which it has moved, e.g. $n = 3$ to $n = 1$. (1)

4 Like electrons in their energy shells, protons and neutrons also occupy energy levels in the nucleus (1). When energy changes occur in the nucleus, high-energy / high-frequency gamma photons are emitted (1) due to greater energy levels occurring in the nucleus. (1)

199. Extended response – Electromagnetic spectrum

*Answer could include the following points: (6)

- A certain microwave frequency will heat water. This heating effect could damage body cells, which are mostly water.
- Infrared waves are used for cooking food. Our skin can absorb infrared waves which we feel as heat. Over-exposure to infrared waves can damage or destroy body cells / cause burns to the skin.
- Over-exposure to sunlight results in damage to body cells from infrared and ultraviolet waves.
- Ultraviolet waves carry high amounts of energy, which can damage body cells.
- Too much exposure to ultraviolet waves can result in skin cancer.
- Ultraviolet radiation can damage eyes.
- X-rays and gamma rays carry very high amounts of energy and can penetrate the body.
- Over-exposure to X-rays and gamma rays can cause mutations in DNA that can kill cells or cause cancer.
- Damage from infrared and ultraviolet waves in sunlight can be reduced by the use of sunscreen.
- Sunglasses provide some protection from ultraviolet waves for those involved in activities where waves are reflected, such as skiing, sailing and flying.

200. Structure of the atom

1 (a) protons labelled in the nucleus (+ charge) (1)

(b) neutrons labelled in nucleus (0 charge) (1)

(c) electrons labelled as orbiting (– charge) (1)

2 (a) The number of positively charged protons in the nucleus (1) is equal to the number of negatively charged electrons orbiting the nucleus. (1)

(b) The atom will become a positively charged ion / charge of +1. (1)

3 (a) A molecule is two or more atoms bonded together. (1) (In the kinetic theory of gases, molecule also describes monatomic gases.)

(b) (i) any pure liquid, e.g. water / H_2O (1)

(ii) any gaseous molecule, e.g. oxygen / O_2 (1)

(iii) any gaseous compound, e.g. carbon dioxide / CO_2 (1)

4 nucleus: 10^{-15} m (1); atom: 10^{-10} m (1); molecule: 10^{-9} m (1)

201. Atoms and isotopes

1 (a) the name given to particles in the nucleus (1)

(b) the number of protons in the nucleus (1)

(c) the total number of protons and neutrons in the nucleus (1)

2 C (1)

3 Isotopes will be neutral because the number of positively charged protons (1) still equals the number of negatively charged electrons. (1)

4 $^{39}_{19}$K: mass number 39 (1); atomic number 19 (1)

5 C (1)

6 They all have 3 protons / they have the same proton number / atomic number (1). The first isotope has 3 neutrons, the second isotope has 4 neutrons and the third isotope has 5 neutrons (1). (all three needed) The first isotope has 6 nucleons, the second isotope has 7 nucleons and the third isotope has 8 nucleons. (1) (all three needed)

202. Atoms, electrons and ions

1 A (1)

2 (a) When an atom absorbs electromagnetic radiation an electron **(1)** moves to a higher energy level. **(1)**

(b) When an atom emits electromagnetic radiation an electron **(1)** moves to a lower energy level. **(1)**

3 (a) Atoms: Li, Cu **(1)**; Ions: F⁻, Na⁺, B⁺, K⁺ **(1)**

(b) Atoms are neutral and have no overall charge **(1)**. Ions have gained (−) or lost (+) an electron / have become negatively or positively charged. **(1)**

4 Two from: an atom can lose an electron by friction (electrostatics) **(1)**. An atom can be made to lose an electron by ionising radiation (radioactivity) **(1)**. By diagram: atom with an electron being removed with force arrow labelled friction **(1)**; atom absorbing a photon with an electron being ejected. **(1)**

203. Ionising radiation

1 B **(1)**

2 alpha: very low, stopped by 10 cm of air; beta minus: low, stopped by thin aluminium; neutron: high; gamma: very high, stopped by very thick lead. All four correct for **3 marks**; two correct for **2 marks**; one correct for **1 mark**.

3 (a) no change **(1)**

(b) high-energy electron **(1)**

(c) moderately ionising **(1)**

4 (a) beta-plus (positron) **(1)**

(b) alpha particle **(1)**

(c) neutron **(1)**

5 Compared to other types of ionising radiation, the chance of collision with air particles at close range is high **(1)** because the alpha particles have a large positive charge / are massive compared to other types of radiation **(1)**. Once an alpha particle has collided with another particle it loses its energy. **(1)**

204. Background radiation

1 Radon is a radioactive element **(1)** that is produced when uranium in rocks decays. **(1)**

2 Levels can vary because of the different rocks that occur naturally in the ground **(1)**. They can also vary due to the use of different rocks such as granite in buildings. **(1)**

3 natural: two from: air, cosmic rays, rocks in the ground, food **(1)**; man-made: two from: nuclear power; medical treatment; nuclear weapons **(1)**

4 (a) south-east 0.27 Bq **(1)**; south-west 0.30 Bq **(1)**

(b) south-west **(1)**

5 (a) As the uranium in rocks decays radon gas seeps out **(1)** from the soil and into homes and buildings. **(1)**

(b) When radon gas is inhaled, the alpha particles can be absorbed by the lungs **(1)** and can be ionising / dangerous in large amounts. **(1)**

205. Measuring radioactivity

1 Photographic film is used by nuclear industry workers, who wear a film badge **(1)** which becomes darker when exposed to radiation **(1)**. This monitors levels of radiation to which the workers are exposed. **(1)**

2 A thin wire is connected to +400 V **(1)**. Atoms of argon are ionised **(1)**. Electrons travel towards the thin wire **(1)**. The amount of radiation detected is shown by the rate meter. **(1)**

3 The student is correct. When radiation is more ionising, it is more likely to create ions **(1)** so more highly ionising radiation is more likely to ionise the argon in the tube **(1)**, which means that it is more likely to cause a current and be recorded on the rate meter. **(1)**

4 Aluminium absorbs beta particles **(1)** and lead absorbs gamma rays **(1)** so this enables the type of radiation to which the wearer has been exposed to be identified. **(1)**

206. Models of the atom

1 The plum pudding model showed the atom as a 'solid', positively charged **(1)** particle containing a distribution of negatively charged electrons **(1)** while the Rutherford model showed the atom as having a tiny, dense, positively charged nucleus **(1)** surrounded by orbiting negatively charged electrons. **(1)**

2 Rutherford fired positively charged alpha particles at atoms of gold foil; most went through showing that there were large spaces in the atom **(1)**. Some were repelled or deflected **(1)** showing that the nucleus was positively charged. **(1)**

3 (a) A **(1)**

(b) The Bohr model showed that electrons **(1)** orbit the atom at different energy levels **(1)** and those electrons have to acquire precise amounts of energy to move up to higher levels **(1)**. The model was an improvement because it was able to explain emission and absorption spectra which enables the atom to be stable. **(1)**

4 (a) Electrons can absorb specific amounts of energy from photons **(1)**. These transfer energy to the electrons which become excited / have more energy so they move to a higher energy level. **(1)**

(b) A photon, of equivalent energy to the energy lost from the 'excited' electron, is emitted from the atom. **(1)**

207. Beta decay

1 Beta-minus decay is when a neutron (n) changes to a proton (p) releasing a high-energy electron (e⁻) **(1)**. Beta-plus decay is when a proton (p) changes to a neutron (n) releasing a high-energy positron (e⁺). **(1)**

2 (a) 7 **(1)**

(b) 12 **(1)**

3 (a) In beta-minus decay, a neutron decays into a proton **(1)** and a high-energy electron (beta-minus particle) is emitted. **(1)**

(b) In beta-plus decay, a proton decays into a neutron **(1)** and a positron (beta-plus particle) is emitted. **(1)**

4 In archaeology, beta decay is used to date objects using radioactive carbon dating **(1)**. In medicine, beta decay is used for producing images in positron (PET) scanning. **(1)**

208. Radioactive decay

1 B and C **(1)**

2 beta-minus, charge (−1) **(1)**; beta-plus, charge (+1) **(1)**

3 The total mass number before a reaction / decay must be the same as the total mass number after a reaction / decay **(1)** so in any nuclear decay the total mass number on either side of the decay equation must balance. **(1)**

4 In neutron decay, a neutron **(1)** is emitted and a new isotope of the element is formed. **(1)**

5 B **(1)**

6 (a) (i) Add 208 to Po **(1)**; alpha **(1)**

(ii) Add 86 to Rn **(1)**; alpha **(1)**

(iii) Add 42 to Ca **(1)**; beta-minus **(1)**

(iv) Add 9 to Be **(1)**; neutron **(1)**

(b) Nucleons often rearrange themselves **(1)** following alpha or beta decay. This causes energy to be emitted as a gamma photon / wave. **(1)**

209. Half-life

1 (a) 8 million nuclei **(1)**

(b) 9.3 ÷ 3.1 = 3 half-lives **(1)**, 1 half-life – 8 million; 2 half-lives – 4 million; 3 half-lives – 2 million nuclei **(1)**

2 The activity is 400 Bq at 1.5 min **(1)** (between 1.3 and 1.7 is allowed). Half this activity is 200 Bq, which is at 6.5 min **(1)** (between 6.3 and 6.7 is allowed) so the half-life is 6.5 min − 1.5 min = 5 min **(1)** (Answers between 4.7 and 5.3 min are allowed). (If you used other points on the graph and found an answer of around 5 min you would get full marks. For this question your working can just be pairs of lines drawn on the graph).

3 The prediction is based on the half-life of caesium **(1)**. You would expect that from 1986 to 2016, radioactivity would have fallen to half this level **(1)**. The level of radioactivity does not fall as rapidly because of background radiation / radioactive materials in the soil. **(1)** (Marks are awarded for discussion of the source of the prediction, i.e. the half-life of caesium, and recognising that other substances will release radiation in addition to the caesium).

210. Dangers of radiation

1 Two from: hospital **(1)**; dental surgery **(1)**; radiography / X-ray department **(1)**; nuclear power plant **(1)**

2 (a) Ionising means that atoms become charged, due to the removal of electrons in this case. **(1)**

(b) Ions in the body can cause damage to cell tissue (1), which can lead to DNA mutations / cancer. (1)

3 (a) Employers can limit the time of exposure (1); workers can wear protective clothing / wear a lead apron (1); increase distance from the source. (1)

(b) The amount of energy / dose of radiation that a person has been exposed to is monitored by wearing a film badge (1). This is checked each day. (1)

4 (a) The intensity of radiation decreases with distance / long handled tongs maximise the distance from the hands / body. (1)

(b) Lead minimises the penetration of gamma waves / stops alpha and beta / radioactive particles / radiation. (1)

5 X-rays carry enough energy (1) to ionise atoms by removing electrons. (1)

211. Contamination and irradiation

1 (a) Radium was useful because it was luminous, allowing watches to be used in the dark. (1)

(b) Before 1920 the effects of radium were not known / recognised (1) so it was thought that it was safe to use (1). It was banned from use once the dangers were known. (1)

2 external contamination: radioactive particles come into contact with skin, hair or clothing; internal contamination: a radioactive source is eaten, drunk or inhaled; irradiation: a person becomes exposed to an external source of ionising radiation. All correct for **2 marks**, 1 correct for **1 mark**.

3 (a) Any suitable example, e.g. contaminated soil may get on to hands. (1)

(b) Any suitable example, e.g. contaminated dust or radon gas may be inhaled. (1)

4 Internal contamination means that the alpha particles come into contact with the body through inhalation or ingestion (1) where they are likely to cause internal tissue damage (1). Alpha particles that are irradiated are less likely to cause damage because they have to travel through air (1) and are therefore less likely to ionise body cells (1) (at distances of over 10 cm).

212. Extended response – Radioactivity

*Answer could include the following points: (6)

- All three types of radiation can pass into / penetrate different materials.
- Alpha particles have high relative mass and so transfer a lot of energy when they collide, so they are good at ionising.
- Alpha particles produce a lot of ions in a short distance, losing energy each time.
- Alpha particles have a short penetration distance so are stopped by low density / thin materials such as a few centimetres of air and a sheet of paper.
- Beta particles have a low relative mass and can pass into / through more materials than alpha particles.
- Beta particles are less ionising than alpha particles and can be stopped by 3 mm thick aluminium.
- Gamma waves are high frequency EM waves and can travel a few kilometres in air.
- Gamma waves are weakly ionising and need thick lead or several metres of concrete to stop them.

213. Work, energy and power

1 D (1)

2 (a) gravitational potential energy store (1)
(b) thermal energy store (1)
(c) chemical energy store (1)

3 energy transferred = 15 000 J, time taken = 20 s; $P = E \div t = 15\,000\,J \div 20\,s$ (1) = 750 (1) W (1)

4 work done = $F \times d = 600\,N \times (20 \times 0.08\,m)$ (1) = 960 (1) J (1)

5 $P = E \div t$ so $t = E \div P$ (1) = 360 000 J $\div$ 200 W (1) = 1800 s (1) (or 30 minutes)

214. Extended response – Energy and forces

*Answer could include the following points: (6)

- This is described as a mechanical process because the turbine is moved by the kinetic energy of the moving air.

- Mechanical energy is the sum of potential and kinetic energy.
- The mechanical energy of the moving air gives the air particles the ability to apply a force and cause a displacement of the blades.
- Mechanical processes become wasteful when they cause a rise in temperature so dissipating energy to the thermal store of the environment.
- Rise in temperature is caused by friction between moving objects / materials.
- It is important to keep friction as low as possible to minimise wasted energy.
- By reducing wasted energy the wind turbines can be made more efficient.
- Higher efficiency will mean more electricity is generated.

215. Interacting forces

1 (a) gravitational (1); magnetic (1); electrostatic (1)
(b) Gravitational fields are different because they only attract (1) whereas magnetic and electrostatic fields attract and repel. (1)

2 A (1) and C (1)

3 Weight is a vector because it has a direction (downwards) (1). Normal contact force is a vector because it has a direction (upwards / opposite to weight). (1)

4 (a) The horizontal contact forces are pull (by the student on the bag) and friction (of the bag against the floor) (1). The forward pull force on the bag is equal to the opposing frictional force (1) (so the bag moves at constant velocity). (1)
(b) weight and normal contact / reaction force (1)

216. Free-body force diagrams

1 B (1)

2 (a) arrow above bird pointing upward (1); arrow below bird pointing downward (1) (both arrows must be the same length).
(b) Reaction force of branch upwards = 20 (1) N (1); weight downwards = 20 (1) N (1)

3 **1 mark** for each arrow. Longest arrow to left indicating resultant acceleration forwards.

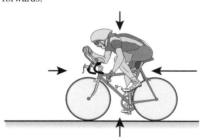

4 (a) 3.6 cm (1)
(b) 7.2 N (1)

217. Resultant forces

1 (a) A: 9.5 N (1); B: 2 N (1); C: 4.5 N (1); D: 12.75 N (1)
(b) A: up (1); B: up (1); C: to the right (1); D: to the left (1)

2 B (1)

3 (a) diagonal arrow: 2.5 cm (1)
(b) 50 N (1)

4 scale correct, e.g. vertical 2 cm (6 N) and horizontal 5 cm (15 N) (1); hypotenuse = 5.4 cm (1) represents 16.2 N (1)

218. Extended response – Forces and their effects

*Answer could include the following points: (6)

- When the drone takes off the downwards force of the rotor blade creates a reaction force (Newton's third law).
- The reaction force is greater than the downwards force of the weight of the drone and it moves upwards.
- The vertical resultant force continues during the flight of the drone.
- The height can be adjusted by increasing or decreasing the upwards reaction force (due to the rotor blades).
- To move to a location, an additional resultant horizontal force is required.

- As the drone flies horizontally, the thrust of the drone must be greater than the air resistance acting in the opposite direction. **(1)**
- When these two are balanced or zero the drone will hover. **(1)**

219. Circuit symbols

1 When there is an electric current in a resistor, there is an energy transfer which heats the resistor. **(1)**

2 (a) C **(1)** and D **(1)**
   (b) (i) The thermistor responds by changing resistance with changes in temperature. **(1)**
       (ii) The LDR responds by changing resistance with changes in light intensity. **(1)**

3

Component	Symbol	Purpose
ammeter	(A)	measures electric current **(1)**
fixed resistor		provides a fixed resistance to the flow of current **(1)**
diode		allows the current to flow one way only **(1)**
switch	or	allows the current to be switched on or off **(1)**

4 Diagram showing series circuit diagram with battery **(1)**; ammeter in series **(1)** and switch in series **(1)**; thermistor in series **(1)**; motor in series **(1)**; voltmeter connected across the motor in parallel. **(1)**

220. Series and parallel circuits

1 (a) series: $A_2 = 3$ A **(1)**; $A_3 = 3$ A **(1)**; parallel: $A_2 = 1$ A **(1)**; $A_3 = 1$ A **(1)**; $A_4 = 1$ A **(1)**
   (b) In a series circuit the current is the same throughout the circuit **(1)**. In a parallel circuit the current splits up in each branch. **(1)**

2 (a) series: $V_2 = 3$ V **(1)**; $V_3 = 3$ V **(1)**; $V_4 = 3$ V **(1)**; parallel: $V_2 = 9$ V **(1)**; $V_3 = 9$ V **(1)**; $V_4 = 9$ V **(1)**
   (b) In a series circuit the potential difference is shared / splits up across the components in the circuit **(1)**. In a parallel circuit the potential difference across each branch is the same as the supply potential difference. **(1)**

3 (a) In a parallel circuit, each component is supplied with sufficient potential difference to work properly **(1)**. If a fault develops, other parts of the circuit will still work. **(1)**
   (b) The lamps would share the potential difference in the circuit so each component would not operate at full capacity **(1)**. If a fault developed, the whole supply would be cut as there would be no other route for the current to take. **(1)**

221. Current and charge

1 (a) An electric current is the rate **(1)** of flow of charge (electrons in a metal). **(1)**
   (b) $Q = I \times t = 4$ A $\times 8$ s **(1)** $= 32$ **(1)** coulombs / C **(1)**

2 (a) (i) $A_1 = 0.3$ A **(1)**
       (ii) $A_3 = 0.3$ A **(1)**
   (b) Add another cell / increase the energy supplied. **(1)**
   (c) The electrons move around the circuit in one continuous path **(1)** so the current leaving the cell is the same as the current returning to it. **(1)**

3 (a) Any series circuit diagram with a component (e.g. lamp) **(1)** and an ammeter. **(1)**
   (b) stopwatch / timer **(1)**

222. Energy and charge

1 Current is the charge flowing per unit time **(1)**. Potential difference is the energy transferred per unit of charge. **(1)**

2 $E = Q \times V$ **(1)** $= 30$ C $\times 9$ V **(1)** $= 270$ J **(1)**

3 $E = Q \times V$ so $Q = E \div V$ **(1)** $= 125$ J $\div 5$ V **(1)** $= 25$ C **(1)**

4 $Q = E \div V = 600$ J $\div 20$ V **(1)** $= 30$ C **(1)**; $Q = I \times t$, so $t = Q \div I = 30$ C $\div 0.15$ A **(1)** $= 200$ s (or 3 min 20 s) **(1)**

223. Ohm's law

1 D **(1)**

2 Ohm's law means that the rate of flow of electrons (the current) flowing through the resistor **(1)** is directly proportional to the potential difference across the resistor. **(1)**

3 (a) $R = V \div I = 12$ V $\div 0.20$ A **(1)** $= 60$ Ω **(1)**
   (b) $R = 22$ V $\div 0.40$ A **(1)** $= 55$ Ω **(1)**
   (c) $R = 9$ V $\div 0.03$ A **(1)** $= 300$ Ω **(1)**
   (d) resistor in part (c) **(1)**

4 (a) Line A: straight line through origin **(1)**; line B: straight line through origin, different gradient. **(1)**
   (b) the line with the lower gradient **(1)**

224. Resistors

1 A **(1)**

2 (a) $20 + 30 + 150$ **(1)** $= 200$ Ω **(1)**
   (b) (i) The sum of the potential differences across the resistors connected in series must equal the potential difference across the battery. **(1)**
       (ii) $R_T = 200$ Ω, $I_T = 0.03$ A; $V = I \times R = 0.03$ A $\times 200$ Ω **(1)** $= 6$ V **(1)**; two identical cells so each cell supplies 3 V **(1)**

3 In a parallel circuit the current is inversely proportional to the resistance (Ohm's law) **(1)**. It will divide up between the available paths **(1)** and more current will travel down the path of lower resistance **(1)**. In this case, using Ohm's law, $I = V \div R$, so $0.9 \div 15 = 0.06$ A and $0.9 \div 10 = 0.09$ A. **(1)**

225. I–V graphs

1 (a) C **(1)**
   (b) As the potential difference increases the current increases **(1)** in a linear / proportional relationship. **(1)**
   (c) As the potential difference increases the current increases **(1)** but the gradient of the line gets less steep / shallower or the increase in current becomes smaller as potential difference continues to increase. **(1)**

2 (a) Fixed resistor: same as graph A in Q1 **(1)**; filament lamp: same as graph B in Q1. **(1)**
   (b) The graphs are a different shape from each other because the fixed resistor (at constant temperature) is ohmic / obeys Ohm's law so the current and potential difference have a proportional relationship **(1)**. The filament lamp does not obey Ohm's law, so the relationship between current and potential difference is not proportional. **(1)**

3 Data can be collected using an ammeter to measure current **(1)** and a voltmeter to measure potential difference **(1)**. A variable resistor **(1)** should be included which will allow different values of current to be obtained **(1)**. Resistance can then be calculated from Ohm's law. **(1)**

226. Electrical circuits

1 (a) Two resistors in same loop **(1)**; at least one ammeter shown and connected in series **(1)** at least one voltmeter shown and connected in parallel across a resistor or cell / battery. **(1)**

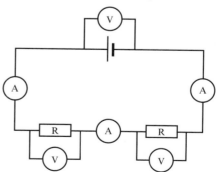

   (b) two resistors in separate loops **(1)**; at least one ammeter shown and connected in main circuit or in a loop connected in series **(1)**; at least one voltmeter shown and connected in parallel across a resistor or cell / battery **(1)**

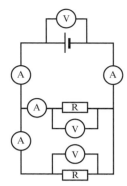

(c) Current is the same at any point in a series circuit (**1**) but will split up at a junction in a parallel circuit (**1**). The sum of the potential difference across components in a series circuit equals the potential difference of the cell (**1**). The sum of the potential difference across components in each loop in a parallel circuit equals the potential difference of the cell. (**1**)

2 (a) Connect a cell, a filament lamp, a variable resistor and an ammeter in a series circuit (**1**). Connect the voltmeter across the filament lamp in parallel (**1**). Adjust the variable resistor setting to obtain a number of readings for current and potential difference. (**1**)

(b) Ohm's law: resistance = potential difference ÷ current ($R = V \div I$) (**1**)

(c) Plotting a graph of I against V shows how the current through it varies with the potential difference across it. The gradient of the I–V graph is equal to $1 / R$ (**1**), so inverting the value of the gradient gives you the resistance, R. (**1**)

3 One from: Resistors can become hot and cause burns (**1**) or fire. (**1**)

227. The LDR and the thermistor

1

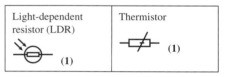

Light-dependent resistor (LDR)	Thermistor
(**1**)	(**1**)

2 D (**1**)

3 (a) The resistance goes down (more current flows) as the light becomes more intense (brighter). (**1**)

(b) The resistance goes down (more current flows) as the temperature goes up. (**1**)

4 The lamp lights up when the temperature is high (**1**) because the current through the lamp and the thermistor will be high when the resistance of the thermistor falls. (**1**)

5 When the level of light increases, the resistance decreases (**1**) and the current increases. (**1**)

228. Current heating effect

1 A (**1**)

2 When a conductor is connected to a potential difference the free electrons (**1**) move through the lattice of metal ions (**1**). As they do so, collisions (**1**) occur where the kinetic energy is transferred into thermal energy (**1**), causing the heating effect.

3 Three suitable examples, e.g. electric fire (**1**), hairdryer (**1**), kettle (**1**), iron (**1**), toaster (**1**)

4 All the appliances draw a certain amount of current (**1**). Those that have heating elements draw more (**1**). Even though the plugs are earthed, too much current being drawn (**1**) could cause a heating effect / fire in the multi-socket extension lead. (**1**)

5 Domestic filament lamps that were only about 5% efficient would have transferred about 95% (**1**) of the electrical energy as 'wasted' thermal energy (**1**). The lamps were withdrawn and replaced with lamps that transferred more useful energy as light. (**1**)

229. Energy and power

1 (a) Using the equation for power $P = I \times V = 5$ A $\times 230$ V (**1**) = 1150 (**1**) W

(b) $E = I \times V \times t = 0.2$ A $\times 4$ V $\times 30$ s (**1**) = 24 (**1**) J (**1**)

2 (a) $P = I \times V$ so $I = P \div V$ (**1**) = 3 W $\div 6$ V (**1**) = 0.5 A (**1**)

(b) $E = I \times V \times t = 0.5$ A $\times 6$ V $\times 300$ s (**1**) = 900 (**1**) J (**1**) (or $E = P \times t = 3$ W $\times 300$ s (**1**) = 900 (**1**) J (**1**))

(c) $P = I^2 \times R$ (**1**) = $(0.5$ A$)^2 \times 240\ \Omega$ (**1**) = 60 W (**1**)

230. A.c. and d.c. circuits

1 (a) An alternating current is an electric current that changes direction regularly (**1**) and its potential difference is constantly changing. (**1**)

(b) A direct current is an electric current in which all the electrons flow in the same direction (**1**) and its potential difference has a constant value. (**1**)

2 (a) $P = E \div t$ so $E = P \times t = 2000 \times (15 \times 60)$ s (**1**) = 1 800 000 J (**1**)

(b) $E = 1500$ W $\times 25$ s (**1**) = 37 500 J (**1**)

(c) $E = 10 \times (6 \times 60 \times 60)$ s (**1**) = 216 000 J (**1**)

3 (a) The current is a direct current (**1**) because the electrons all flow in the same direction. (**1**)

(b) There should be one horizontal line anywhere on the screen. (**1**)

231. Mains electricity and the plug

1 (a) earth wire (green and yellow) (**1**); live wire (brown) (**1**); neutral wire (blue) (**1**); fuse (**1**)

(b) The fuse is connected to the live wire (**1**) because it carries the current into the appliance. (**1**)

2 brown: Electrical current enters the appliance at 230 V (**1**). blue: Electrical current leaves the appliance at 0 V through this wire (**1**). green / yellow: This is a safety feature connected to the metal casing of the appliance. (**1**)

3 When a large current enters the live wire (**1**) this produces thermal energy (**1**) which melts the wire in the fuse (**1**). The circuit is then broken. (**1**)

4 (a) When a current is too high (**1**) a strong magnetic field is generated which opens a switch (held back by a spring) (**1**). This 'breaks' the circuit (**1**), making it safe.

(b) The earth wire is connected to the metal casing (**1**). If the live wire becomes loose and touches anything metallic (**1**) the current passes through the earth wire. (**1**)

232. Extended response – Electricity and circuits

*Answer could include the following points: (**6**)

- The thermistor can be connected in series with an ammeter to measure current with a voltmeter connected in parallel across it to measure potential difference.
- Ohm's law can be referred to in calculating the resistance.
- When the temperature is low the resistance of the thermistor will be high, allowing only a small current to flow.
- When the temperature is high the resistance of the thermistor will be low, allowing a larger current to flow.
- The light-dependent resistor can be connected in series with an ammeter to measure current with a voltmeter connected in parallel across it to measure potential difference.
- When light levels are low (dark) the resistance of the light-dependent resistor will be high, allowing only a small current to flow.
- When light levels are high (bright) the resistance of the light-dependent resistor will be low, allowing a larger current to flow.
- Thermistors can be used in fire alarms as a temperature sensor to switch on an alarm.
- Light-dependent resistors can be used in security systems as a light sensor to switch on a light.

233. Magnets and magnetic fields

1 (a) field lines out (arrows) at N (**1**); field lines in (arrows) at S (**1**); field lines close at poles (**1**), further apart at sides (**1**)

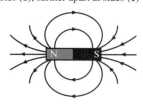

(b) uniform field: parallel field lines **(1)**; arrows from N to S **(1)**

2 Both a bar magnet and the Earth have north and south poles **(1)**. They also both have similar magnetic field patterns. **(1)**

3 A temporary magnet is used for an electric doorbell because it can be magnetised when the current is switched on **(1)**, which attracts the soft iron armature to ring the bell **(1)**, and de-magnetised when the current is switched off **(1)** (returning the armature away from the bell).

4 Rajesh can do a second test by moving a permanent magnet near the magnetic materials **(1)**. Those that are attracted but not repelled will be temporary magnets **(1)**. The materials that can be attracted and repelled are permanent magnets. **(1)**

234. Current and magnetism

1 (a) at least two concentric circles on each diagram **(2)**
(b) anticlockwise arrows on dot diagram **(1)**; clockwise arrows on cross diagram **(1)**

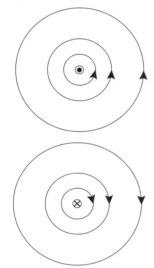

2 B **(1)**

3 (a) The strength of the magnetic field depends on the size of the current in the wire **(1)** and the distance from the wire. **(1)**
(b) (i) x-axis marked 'Current' **(1)**
(ii) x-axis marked 'Distance from the wire' **(1)**

4 (a) (i) 2*B* **(1)**
(ii) 0.5*B* **(1)**
(b) (i) 0.5*B* **(1)**
(ii) 2*B* **(1)**

235. Current, magnetism and force

1 D **(1)**

2 (a) The force occurs because a magnetic field is generated by the current in the wire **(1)**; this interacts with the magnetic field between the two magnets **(1)**, creating an equal and opposite force. **(1)**
(b) The force is at a maximum when the wire is at right angles to the magnetic field. **(1)**

3 First finger – field **(1)**; SeCond finger – Current **(1)**; ThuMb – Movement **(1)**

4 The size of the force can be increased by increasing the strength of the magnetic field **(1)** or by increasing the current. **(1)**

5 $F = B \times I \times \ell = 0.0005\,T \times 1.4\,A \times 0.30\,m$ **(1)** = 0.000 21 or 2.1×10^{-4} **(1)** N / newtons **(1)**

236. Extended response – Magnetism and the motor effect

*Answer could include the following points: **(6)**

- A long straight conductor could be connected to a cell, an ammeter and a small resistor to prevent overheating in the conductor.
- When the current is switched on the direction of the current generated around a long straight conductor can be found using the right hand grip rule (shown by the thumb).
- When the current is switched on the direction of the magnetic field generated around a long straight conductor can be found using the right hand grip rule (shown by the fingers).
- The right hand grip rule points the thumb in the direction of conventional current and the direction of the fingers show the direction of the magnetic field.
- A card can be cut halfway through and placed at right angles to the long straight conductor. A plotting compass can be used to show the shape and direction of the magnetic field.
- The shape of the magnetic field around the long straight conductor will be circular / concentric circles as the current flows through it.
- The strength of the magnetic field depends on the distance from the conductor.
- The concentric magnetic field lines mean that the field becomes weaker with increasing distance.
- The strength of the magnetic field can be increased by increasing the current.

237. Electromagnetic induction and transformers

1 (a) Move the wire up (or down) through a magnetic field. **(1)**
(b) Move the wire in the opposite direction to (a) OR turn the magnet around but move the wire in the same direction as before. **(1)**
(c) Three from: move the wire faster **(1)**; use a stronger magnet **(1)**; use thicker wire / more lengths of wire **(1)**; use more loops / turns in the wire **(1)**; wind the wire around an iron core **(1)**

2 $V_p \times I_p = V_s \times I_s$, so $V_p = 110\,V$, $V_s = 12\,V$, $I_s = 1.5\,A$ so $I_p = V_s \times I_s / V_p$ **(1)** so $I_p = 12 \times 1.5 / 110$ **(1)** = 0.16 **(1)** A

3 (a) A voltage in one coil of wire causes / induces a voltage **(1)** in the second, unconnected coil in a transformer. **(1)**
(b) Two (separate) coils of wire wound onto an iron core **(1)**; no electrical connection between the two coils of wire **(1)**. Soft iron is used as the core because it allows the magnetic field to be reversed rapidly. **(1)**
(c) $V_p \times I_p = V_s \times I_s$ so $(V_p \times I_p) / V_s = I_s = (2500 \times 20) / 200\,000$ **(1)** = 0.25 A **(1)**

238. Transmitting electricity

1 The voltage is increased so that the current goes down **(1)**, which reduces the heating effect due to resistance **(1)** and means less energy is wasted in transmission. **(1)**

2 step-down transformer – decreases voltage **(1)**

3 $P = IV = 20\,000\,A \times 25\,000\,V$ **(1)** = 500 000 kW **(1)**

4 Step-up transformers are used to increase the voltage **(1)** as it leaves the power station for transmission through the National Grid **(1)**. Near homes, step-down transformers are used to reduce the voltage **(1)** to make it safer for domestic use. **(1)**

239. Extended response – Electromagnetic induction

*Answer could include the following points: **(6)**

- Electromagnetic induction means that a voltage in one coil causes / induces a voltage in the second coil.
- Transformers are used in the National Grid to increase the voltage to reduce the amount of wasted energy in transmission lines.
- If the voltage is increased, the current is decreased.
- Smaller current means that less energy is wasted by heating / efficiency is improved.
- Factors affecting the size and direction of an induced potential difference / voltage include: the voltage in the primary circuit, the number of turns in the primary coil, the number of turns in the secondary coil, how fast the magnetic field changes direction.
- Transformers only work with an alternating current in the primary coil which creates a continuously changing magnetic field in the iron core of the transformer.

- The changing magnetic field in the core induces a changing potential difference / voltage in the secondary coil, according to the number of coils in the secondary coil.

240. Changes of state

1 All four links correct for **2 marks**, two correct for **1 mark**. particles move around each other – liquid – some intermolecular forces; particles cannot move freely – solid – strong intermolecular forces; particles move randomly – gas – almost no intermolecular forces

2 (a) They all consist of particles. **(1)**
   (b) Particles have different amounts in the kinetic energy store **(1)** and experience different intermolecular forces. **(1)**

3 B **(1)**

4 At boiling point the liquid changes state **(1)** so the energy applied after boiling point is reached goes into breaking bonds **(1)** between the liquid particles. The particles gain more energy and become a gas. **(1)**

5 The kinetic energy **(1)** of the particles decreases **(1)** as the ice continues to lose energy to the surroundings. This is measured as a fall in temperature. **(1)**

241. Density

1 $\rho = m \div V$ **(1)** $= 4000$ g $\div 5000$ cm^3 **(1)** $= 0.8$ g / cm^3 **(1)**

2 A **(1)**; B **(1)**

3 volume $= 10$ cm $\times 25$ cm $\times 15$ cm $= 3750$ cm^3; $\rho = m \div V$ so $m = \rho \times V = 3$ g / cm$^3 \times 3750$ cm^3 **(1)** $= 11\,250$ g **(1)** $= 11.25$ kg **(1)**

4 Marco has approached this problem by stating a scientific principle **(1)** relating density to states of matter but he has not tried to investigate this **(1)**. Ella has approached this problem by observing and comparing **(1)** different densities, but she has not tried to explain this **(1)**. Both students should expand their approach so that observations are explained through scientific principles. **(1)**

242. Investigating density

1 (a) mass **(1)**
   (b) electronic balance **(1)**

2 (a) The volume of mass may be found by measuring its dimensions **(1)** or by using a displacement method, such as immersing the object in water in a measuring cylinder, to measure how much liquid the mass displaces. **(1)**
   (b) The measurement method is suitable for regular-shaped objects **(1)** whereas the displacement method is best for irregular-shaped objects, where measuring dimensions would be more difficult. **(1)**

3 (a) Place a measuring cylinder on a balance and then zero the scales with no liquid present in the measuring cylinder **(1)**. Add the liquid to the required level **(1)**. Record the mass of the liquid (in g) from the balance and its volume (in cm^3) from the measuring cylinder. **(1)**
   (b) Take the value at the bottom of the meniscus **(1)**. Make sure the reading is taken with the line of sight from the eye to the meniscus perpendicular to the scale to avoid a parallax error. **(1)**
   (c) density = mass ÷ volume = 121 g ÷ 205 cm^3 = 0.59 **(1)** g / cm^3 **(1)**
   (d) One from: Care should be taken when placing liquid on an electronic balance to avoid the risk of electrical shock from wet hands or spillage when using electricity **(1)** – keep hands and working area dry **(1)** OR Care should be taken to avoid spillages to avoid the risk of slippage **(1)** – wipe spills and warn others **(1)** OR Do not use toxic or harmful liquids **(1)** – always check the hazard label on a liquid container. **(1)**

243. Energy and changes of state

1 Specific heat capacity is a measure of how much energy is required to change the temperature of a mass of 1 kg by 1 °C. **(1)**

2 $\Delta Q = m \times c \times \Delta T$ **(1)** $= 0.8$ kg $\times 4200$ J / kg °C $\times 50$ °C **(1)** $= 168\,000$ J **(1)**

3 $Q = m \times L = 25$ kg $\times 336\,000$ J / kg **(1)** $= 8\,400\,000$ J **(1)**

4 (a) melting: added to lower horizontal line; boiling: added to higher horizontal line **(1)** (both needed for the mark)
   (b) The energy being transferred to the material is breaking bonds **(1)** and as a result the material undergoes a phase change. **(1)**

5 $c = \Delta Q \div (m \times \Delta T)$; $\Delta Q_{in} = I \times V \times t = 2.4$ A $\times 12$ V $\times (9 \times 60)$ s **(1)** $= 15\,552$ J **(1)**; $\Delta Q_{in} = \Delta Q_{out}$ so $c = 15\,552 \div (0.8$ kg $\times 25$ °C$)$ **(1)** $= 15\,552 \div 20 = 777.6$ J / kg °C **(1)**

244. Thermal properties of water

1 (a) the amount of energy required to raise the temperature of 1 kg of material by 1 K (or 1 °C) **(1)**
   (b) specific heat capacity = change in thermal energy ÷ (mass × change in temperature) or ($c = \Delta Q \div (m \times \Delta T)$) **(1)**

2 (a) Place a beaker on a balance, zero the balance and add a measured mass of water **(1)**. Take a start reading of the temperature **(1)**. Place the electrical heater into the water and switch on **(1)**. Take a temperature reading every 30 seconds **(1)** until the water reaches the required temperature. **(1)**
   (b) Measure the current supplied, the potential difference across the heater and the time for which the current is switched on **(1)**. Use these values to calculate the thermal energy supplied using the equation $Q = V \times I \times t$. **(1)**
   (c) Add insulation around the beaker **(1)** so less thermal energy is transferred to the surroundings and a more accurate value for the specific heat capacity of the water may be obtained. **(1)**

3 Plot a graph of temperature against time **(1)**. The changes of state are shown when the graph is horizontal (the temperature is not increasing). **(1)**

4 Both experiments use an electrical heater close to water so there is a danger of electric shock – keep electrical wires and switches dry **(1)**. Both experiments use water that could be spilled and cause slippage – report and wipe up immediately. **(1)** (Note: specific latent heat experiments tend not to use glass beakers (which could break and cause cuts in the specific heat capacity experiment) but tend to use metal containers, so glass is not necessarily common to both experiments. The specific heat capacity experiment does not require water to be heated to a level to cause scalds so the hot water / water vapour hazard in the specific latent heat experiment is not necessarily common to both experiments.)

245. Pressure and temperature

1 Temperature is a measurement of the average kinetic energy of the particles in a material. **(1)**

2 (a) 273 K → 0 °C **(1)**; 255 K → −18 °C **(1)**; 373 K → 100 °C **(1)**
   (b) (i) At absolute zero, the volume / pressure **(1)** and kinetic energy of the particles **(1)** of a substance will be zero. **(1)**
       (ii) −273 °C **(1)**

3 (a) As the temperature increases the particles will move faster **(1)** because they gain more energy. **(1)**
   (b) As the particles are moving faster they will collide with the container walls more often **(1)** therefore increasing the pressure. **(1)**
   (c) It increases. **(1)**

4 The average kinetic energy of the particles will also increase by a factor of four **(1)** because temperature and average kinetic energy are directly proportional. **(1)**

246. Extended response – Particle model

*Answer could include the following points: **(6)**

- Solid, liquid and gas states of matter have increasing amounts of kinetic energy of particles (solid to liquid to gas).
- Thermal energy input or output will result in changes to the thermal energy store of the system and will result in changes of state or a change in temperature.
- Changes in states of matter are reversible because the material recovers its original properties if the change is reversed.
- Thermal energy input does not always result in a temperature rise if the energy is used to make or break bonds between particles / result in a change of state.
- Latent heat is the amount of heat / thermal energy required by a substance to undergo a change of state.
- The thermal energy required to change from solid / ice to water (accept converse) is called the latent heat of fusion and is calculated using $Q_f = mL$.
- The thermal energy required to change from liquid to gas / water to steam (accept converse) is called the latent heat of vaporisation and is calculated using $Q_v = mL$.

247. Elastic and inelastic distortion

1 push forces (towards each other) – compression **(1)**; pull forces (away from each other) – stretching **(1)**; clockwise and anticlockwise – bending **(1)**

2 (a) washing line (or any valid example) **(1)**
(b) G-clamp, pliers (or any valid example) **(1)**
(c) fishing rod (with a fish on the line) (or any valid example) **(1)**
(d) dented can or deformed spring (or any valid example) **(1)**

3 After testing, beam 1 would return to the same size and shape as prior to the test **(1)** under the load but would be intact **(1)**. Beam 2 would distort and change shape **(1)** but would (probably) still be intact. **(1)**

4 Car manufacturers install crumple zones / seat belts / airbags **(1)** in cars. These are parts of the car body that are designed to distort / change shape **(1)** in the event of a crash. They extend the time taken for a body to come to rest, reducing the force on the body. **(1)**

248. Springs

1 Elastic means that the object will return to original size and shape **(1)** (*both needed for mark*) after the deforming force is removed. **(1)**

2 extension = 0.07 m − 0.03 m = 0.04 m; force = spring constant / $k \times$ extension = 80 N $\times$ 0.04 m **(1)** = 3.2 **(1)** N **(1)**

3 D **(1)**

4 (a) $F = k \times x$ so $k = F \div x$ **(1)** = 30 N $\div$ 0.15 m **(1)** = 200 N / m **(1)**
(b) $E = \frac{1}{2} \times k \times x^2 = \frac{1}{2} \times 200$ N / m $\times (0.15$ m$)^2$ **(1)** = 2.25 J **(1)**

249. Forces and springs

1 (a) Hang a spring from a clamp attached to a retort stand and measure the length before any masses or weights are added using a half-metre ruler, marked in mm **(1)**. Carefully add the first mass or weight and measure the total length of the extended spring **(1)**. Unload the mass or weight and re-measure the spring to make sure that the original length has not changed **(1)**. Add at least five masses or weights and repeat the measurements each time. **(1)**
(b) The elastic potential energy can all be recovered **(1)** and is not transferred to cause a permanent change of shape in the spring. **(1)**

(c) Masses must be converted to force (N) by using $W = m \times g$ / $F = m \times g$ **(1)**. The extension of the spring must be calculated for each force by taking away the original length of the spring from each reading **(1)**. Extension measurements should be converted to metres. **(1)**
(d) (i) The area under the graph equals the work done / the energy stored in the spring as elastic potential energy. **(1)**
(ii) The gradient of the linear part of the force–extension graph gives the spring constant k. **(1)**
(e) Hooke's law **(1)**
(f) energy stored = $\frac{1}{2} \times k \times x^2$ **(1)**

2 The length of a spring is measured with no force applied to the spring whereas the extension of a spring is the length of the spring measured under load / force less the original length. **(1)**

250. Extended response – Forces and matter

*Answer could include the following points: **(6)**

- Two or more forces are required to cause an object to distort / deform / change shape as one force has to hold the object in position.
- Elastic distortion results in the object returning to its original shape whilst inelastic distortion results in a permanent change of shape.
- Energy transferred to the object by a force can be stored if it deforms elastically and the force maintained. When the force is removed, energy may be recovered from the potential elastic / spring energy store of the object.
- When an object permanently changes shape, the energy transferred to the object by a force cannot be recovered as it has, instead, caused permanent distortion / change of shape.
- Metal springs usually exhibit elastic distortion when the force is below that which would cause a permanent change of shape.
- The relationship between force and the extension of a spring is usually directly proportional (if the force doubles, the extension doubles) during elastic distortion / change of shape.
- Elastic distortion may be useful in sports events, such as diving, archery or pole vaulting, where the potential energy stored can be recovered and used by the athlete (any valid example).
- Inelastic distortion may be useful when producing industrial goods, such as car panels or plastic bottles, where a force results in a permanent change of shape (any valid example).

The Periodic Table of the Elements

Key

| relative atomic mass |
| **atomic symbol** |
| name |
| atomic (proton) number |

| 1 |
| **H** |
| hydrogen |
| 1 |

Group 1	Group 2												Group 3	Group 4	Group 5	Group 6	Group 7	Group 0
																		4 **He** helium 2
7 **Li** lithium 3	9 **Be** beryllium 4												11 **B** boron 5	12 **C** carbon 6	14 **N** nitrogen 7	16 **O** oxygen 8	19 **F** fluorine 9	20 **Ne** neon 10
23 **Na** sodium 11	24 **Mg** magnesium 12												27 **Al** aluminium 13	28 **Si** silicon 14	31 **P** phosphorus 15	32 **S** sulfur 16	35.5 **Cl** chlorine 17	40 **Ar** argon 18
39 **K** potassium 19	40 **Ca** calcium 20	45 **Sc** scandium 21	48 **Ti** titanium 22	51 **V** vanadium 23	52 **Cr** chromium 24	55 **Mn** manganese 25	56 **Fe** iron 26	59 **Co** cobalt 27	59 **Ni** nickel 28	63.5 **Cu** copper 29	65 **Zn** zinc 30		70 **Ga** gallium 31	73 **Ge** germanium 32	75 **As** arsenic 33	79 **Se** selenium 34	80 **Br** bromine 35	84 **Kr** krypton 36
85 **Rb** rubidium 37	88 **Sr** strontium 38	89 **Y** yttrium 39	91 **Zr** zirconium 40	93 **Nb** niobium 41	96 **Mo** molybdenum 42	[98] **Tc** technetium 43	101 **Ru** ruthenium 44	103 **Rh** rhodium 45	106 **Pd** palladium 46	108 **Ag** silver 47	112 **Cd** cadmium 48		115 **In** indium 49	119 **Sn** tin 50	122 **Sb** antimony 51	128 **Te** tellurium 52	127 **I** iodine 53	131 **Xe** xenon 54
133 **Cs** caesium 55	137 **Ba** barium 56	139 **La*** lanthanum 57	178 **Hf** hafnium 72	181 **Ta** tantalum 73	184 **W** tungsten 74	186 **Re** rhenium 75	190 **Os** osmium 76	192 **Ir** iridium 77	195 **Pt** platinum 78	197 **Au** gold 79	201 **Hg** mercury 80		204 **Tl** thallium 81	207 **Pb** lead 82	209 **Bi** bismuth 83	[209] **Po** polonium 84	[210] **At** astatine 85	[222] **Rn** radon 86
[223] **Fr** francium 87	[226] **Ra** radium 88	[227] **Ac*** actinium 89	[261] **Rf** rutherfordium 104	[262] **Db** dubnium 105	[266] **Sg** seaborgium 106	[264] **Bh** bohrium 107	[277] **Hs** hassium 108	[268] **Mt** meitnerium 109	[271] **Ds** darmstadtium 110	[272] **Rg** roentgenium 111								

Elements with atomic numbers 112–116 have been reported but not fully authenticated

*The lanthanoids (atomic numbers 58–71) and the actinoids (atomic numbers 90–103) have been omitted.

Physics Equations List

(final velocity)^2 – (initial velocity)^2 = 2 × acceleration × distance

$v^2 - u^2 = 2 \times a \times x$

force = change in momentum ÷ time

$F = \dfrac{(mv - mu)}{t}$

energy transferred = current × potential difference × time

$E = I \times V \times t$

force on a conductor at right angles to a magnetic field carrying a current = magnetic flux density × current × length

$F = B \times I \times l$

potential difference across primary coil × current in primary coil = potential difference across secondary coil × current in secondary coil

$V_p \times I_p = V_s \times I_s$

change in thermal energy = mass × specific heat capacity × change in temperature

$\Delta Q = m \times c \times \Delta\theta$

thermal energy for a change of state = mass × specific latent heat

$Q = m \times L$

energy transferred in stretching = 0.5 × spring constant × (extension)^2

$E = \frac{1}{2} \times k \times x^2$

Notes

Notes

Notes

Notes

Notes

Notes

Published by Pearson Education Limited, 80 Strand, London, WC2R 0RL.

www.pearsonschoolsandfecolleges.co.uk

Copies of official specifications for all Pearson qualifications may be found on the website: qualifications.pearson.com

Text and illustrations © Pearson Education Limited 2017
Typeset and produced by Phoenix Photosetting
Illustrated by Phoenix Photosetting
Cover illustration by Miriam Sturdee

The rights of Stephen Hoare, Nigel Saunders and Catherine Wilson to be identified as authors of this work have been asserted by them in accordance with the Copyright, Designs and Patents Act 1988.

First published 2017

21
10 9 8 7

British Library Cataloguing in Publication Data
A catalogue record for this book is available from the British Library

ISBN 978 1 292 13158 0

Printed in Great Britain by Bell & Bain Ltd, Glasgow

Acknowledgements
Content written by Ian Roberts, Damian Riddle, Julia Salter and Stephen Winrow-Campbell is included in this book. The publishers would also like to acknowledge the contribution of Allison Court.

The authors and publisher would like to thank the following individuals and organisations for permission to reproduce copyright material:

Photographs
Alamy Stock Photo: ALANDAWSONPHOTOGRAPHY 190, Bruce Boulton.co.uk 162; © **Crown copyright:** Contains public sector information licensed under the Open Government Licence v3.0 180; **NASA:** 174; **Pearson Education Ltd:** Oxford Designers & Illustrators Ltd 150; **Science Photo Library Ltd:** BioPhoto Associates 3, Steve Gschmeissner 6, 13

Figures
Graph on page 42: data from National Health Service of the United Kingdom, http://www.pbs.org/wgbh/nova/body/autism-vaccine-myth.html

All other images © Pearson Education

Notes from the publisher

1. In order to ensure that this resource offers high-quality support for the associated Pearson qualification, it has been through a review process by the awarding body. This process confirms that this resource fully covers the teaching and learning content of the specification or part of a specification at which it is aimed. It also confirms that it demonstrates an appropriate balance between the development of subject skills, knowledge and understanding, in addition to preparation for assessment.

Endorsement does not cover any guidance on assessment activities or processes (e.g. practice questions or advice on how to answer assessment questions), included in the resource nor does it prescribe any particular approach to the teaching or delivery of a related course.

While the publishers have made every attempt to ensure that advice on the qualification and its assessment is accurate, the official specification and associated assessment guidance materials are the only authoritative source of information and should always be referred to for definitive guidance.

Pearson examiners have not contributed to any sections in this resource relevant to examination papers for which they have responsibility.

Examiners will not use endorsed resources as a source of material for any assessment set by Pearson.

Endorsement of a resource does not mean that the resource is required to achieve this Pearson qualification, nor does it mean that it is the only suitable material available to support the qualification, and any resource lists produced by the awarding body shall include this and other appropriate resources.

2. Pearson has robust editorial processes, including answer and fact checks, to ensure the accuracy of the content in this publication, and every effort is made to ensure this publication is free of errors. We are, however, only human, and occasionally errors do occur. Pearson is not liable for any misunderstandings that arise as a result of errors in this publication, but it is our priority to ensure that the content is accurate. If you spot an error, please do contact us at resourcescorrections@pearson.com so we can make sure it is corrected.